Taunton's

TRIM COMPLETE

EXPERT ADVICE FROM START TO FINISH

The Taunton Press, Inc.,
63 South Main Street, PO Box 5506,
Newtown, CT 06470-5506
e-mail: tp@taunton.com

Editors: Helen Albert, Jim Tolpin
Copy editor: Candace B. Levy
Indexer: Jay Kreider
Jacket/Cover design: Kimberly Adis
Interior design: Kimberly Adis
Layout: Kimberly Adis
Illustrator: Trevor Johnston
Photographer: Craig Wester, except as noted.

Library of Congress Cataloging-in-Publication Data
Kossow, Greg.
Trim complete / Greg Kossow.
p. cm.
ISBN 978-1-56158-869-5
1. Trim carpentry. I. Title.

TH5695 .K67
694'.6--dc22
2007014399

Printed in the United States of America
10 9 8 7 6 5 4

Construction is inherently dangerous. Using hand or power tools improperly or ignoring safety practices can lead to permanent injury or even death. Don't try to perform operations you learn about here (or elsewhere) unless you're certain they are safe for you. If something about an operation doesn't feel right, don't do it. Look for another way. We want you to enjoy working on your home, so please keep safety foremost in your mind whenever you're in the shop.

I want to thank my wife Lianne for her continued support and patience during this process. Thanks to Jim Tolpin for his continued support and editing and to photographer Craig Wester. The crew at Taunton were the best. Helen Albert's editing and guidance were essential. Jennifer Peters took care of everything I asked of her quickly and efficiently.

I would like to thank the local builders who allowed me to work and photograph their projects and were willing to adjust their schedules to accommodate my needs. Kurt Schweizer at Schweizer Construction, Neil Cavette at Seven Bridges Builders, Vern Garrison at Garrison Construction, and Larry Reichert at Reichert Construction were all a great help in allowing me to photo works in progress. I would like to thank Sebastian Eggert at The Maizefield Company for making time to share his extensive expertise on moldings. Thanks also to finish carpenter Dan Clemmons for sharing his knowledge.

The following suppliers supplied tools and materials for this project: Kreg Tool Company, Bosch Tool Company, Panasonic Tool Company, Makita Tools, Fast Cap, Thomas Compressors, Veto Pro Pac, Collins Tool Company.

contents

>> >> >> >>

contents

DOORS 86

RUNNING MOLDINGS 114

WAINSCOT 144

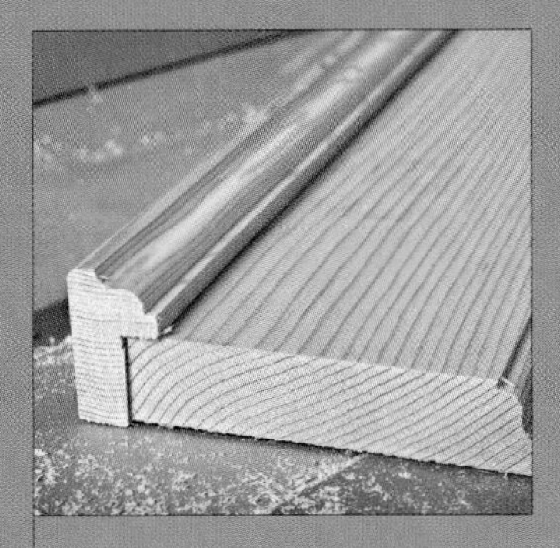

BEFORE YOU START

GOOD PLANNING MEANS FEWER SURprises and problems, so take the time to consider these fundamental questions so you can get going on to the fun stuff!

What tools and skills will I need?

The most important skill you'll need is a working knowledge of your tools. Don't be afraid to gain that knowledge by using up some material practicing the techniques you're going to need.

Become familiar with the tool you'll be using. Practice if necessary. Always make test cuts for repetitive dimensions before before cutting all of your material.

Be aware that installing trim level, even, and with close-fitting joints does require working to closer tolerances than you may be used to in general carpentry—from the first steps of taking and transferring measurements and determining angles to cutting the individual components to the final steps of installing the pieces securely in place.

Aside from carpentry skills, it is also important to become proficient at keeping the materials and procedures organized so you can work efficiently and with a minimum of material-wasting mistakes.

How long will the project take?

While only your own personal experience and level of expertise can determine how quickly you will get through the various tasks involved in the project, the rule of thumb for estimating the overall time frame goes like this: Come up with the best estimate you can muster based on your experience and then double it. If you have never done anything quite like what you are attempting, it may take all that time and more to complete the project. Hopefully, you might surprise yourself and finish early. In any case, though, avoid trying to hurry through the process—otherwise both you and the quality of the finished project are likely going to suffer.

How do I figure out the cost of the project?

It's important to do your homework beforehand to determine the fixed costs of the project before you begin. Draw up a complete materials list. Be sure to account for some waste—the rule of thumb for trim stock is to add to the lineal count by about 15%. Don't forget to add in

PLANNING TRIM FOR BEST APPEARANCE

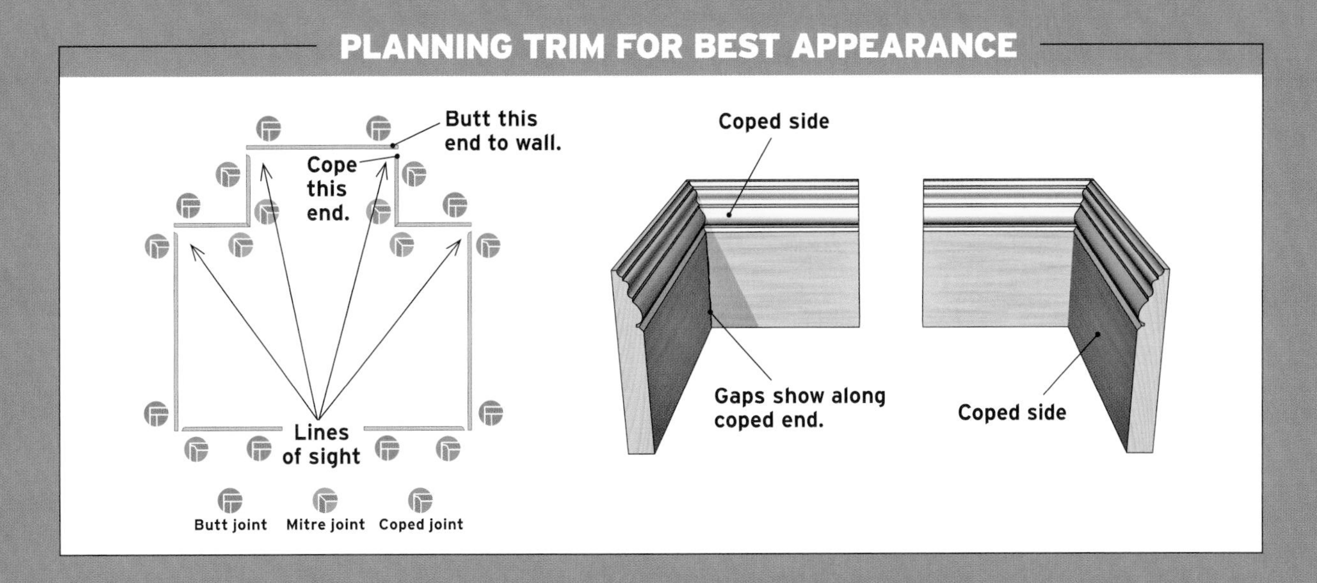

supplies such as sandpaper and glue if you really want a true picture of material costs. Millwork stock varies greatly in price depending on its species, quality, and style. What you choose to go with here will have a significant impact on the final price of the project.

How do I organize the project efficiently?

Before embarking on any project, take into account how the project will fit into the daily life of your family—it will not go very quickly or sanely if you find yourself doing trim work during holidays or other special events.

Begin the planning of the project by deciding the order of the various tasks involved. If the millwork is going to be painted or clear-finished, you must decide whether to prefinish the material before installation. It's not an easy decision: while finishing materials are much easier (and therefore much faster) to apply when the material is not on the walls, doing so takes away your ability to fine-tune the joints by sanding or planning—and that inevitably slows down the installation process.

Another consideration affecting the scope of the entire job is to be sure that windows, doors, and other room elements such as built-ins are ready to accept their trim. You obviously cannot, for examples, trim out doors until the doorjambs are installed or put on window trim until the jamb extensions are in.

What kind of problems could I encounter?

Don't be surprised to find walls and floors disturbingly out of plumb or level; adjacent windows out of alignment with one another; or doors and windows out of plumb. In many cases, you'll find that making small adjustments will correct—at least visually—large problems: Someone would have to look closely at the reveals in order to see their irregularity where as the tapered gap of an out-of-plumb casing running next to a plumb wall or cabinet is more readily noticed.

Planning installation

Some trim carpenters always work from the top down: first installing the ceiling trim work, then the ceiling-to-wall trim, then wall trim, then the wall-to-floor trim. The advantage of this arrangement is that the completed work is always above the work in process and therefore out of harms way.

But it's OK to start with a procedure that you feel more comfortable with in order to build up confidence and sharpen your skills before tackling the more complicated tasks. Just be sure to think it through beforehand to be sure that the order in which you do things won't create complications and conflicts further down the road.

Another way to order the installation is to complete all of one type at a time, such as installing all the baseboard. The advantage to this approach is that you only need to set up the tooling and staging for dealing with a particular trim molding once.

When installing running moldings, good planning means both fewer joints and a better appearance. Consider the line of sight as you enter a room and plan the copes on inside corners to hide any irregularities.

ABOUT YOUR SAFETY

WHILE CUTTING AND INSTALLING trim one is exposed continuously to potentially dangerous situations. Using the proper safety gear, however, can make the difference between getting a warning and getting hurt. But you also to have to use your head: Safety guards on tools cannot always prevent injury. If you are tired and not alert you must learn to recognize when to stop. Make it a rule to avoid working with power or sharp handtools under these conditions. Also avoid attempting something too complicated or potentially dangerous at the end of the workday when you are tired. Finally, make a practice of using appropriate personal safety equipment:

Safety glasses—I have pieces of wood and deflected nails bounce off my glasses too many times to remember.

Hearing protection—I started wearing them too late in my career and my hearing has suffered for it. Peltor makes the best that I have found

Dust mask—Any kind of dust has been proven to be hazardous to your health. Try to minimize the amount of dust being blown around by using on-site portable dust collection system. In situations where the dust cannot be effectively controlled, use a mask or better yet a powered filter that blows filtered air across your face.

Disposable gloves—when using epoxy and CA glues and applying finishes.

In many of the photos in this book, the guards have been removed for clarity. ALWAYS use appropriate guards [regular] when using power tools, as well as featherboards to hold stock, push sticks to keep your hands from the blade and outfeed support.

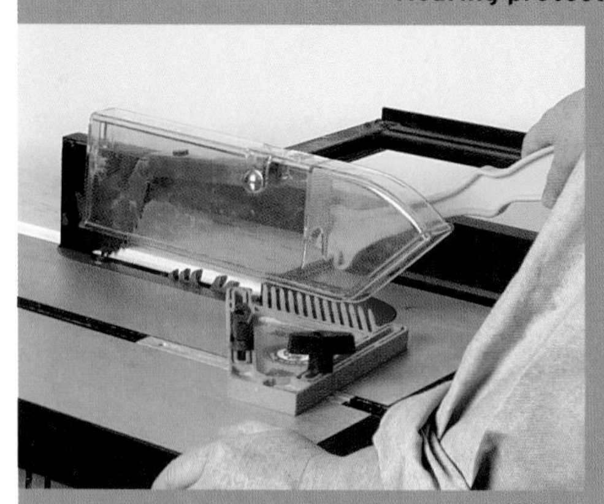

A place to work

A proper place for working and something secure to work on will dramatically improve the quality and speed of your work as well as making it a more pleasurable experience. At the minimum, provide yourself with dedicated stations for the chopsaw and tablesaw and a bench for assembly and hand tool work. A sheet of plywood on a set of saw horses makes an acceptable assembly table while an old solid core door works well for a portable bench top and provides a solid platform for mounting a vice. Ideally, your workplace will be well lit and ventilated in order to deal with the inevitable sawdust.

The minimum sized workspace should be big enough to allow you to move any necessary materials freely around the space without having to move any of the tools or work stations. A good tactic in small spaces is to orient your saw stations to existing door or window openings to allow any long stock to pass through them.

Your workspace also needs to be kept relatively warm and dry. If the floor is concrete be especially vigilant to keep the materials off the floor—or to at least put down a layer of plastic before stocking the material. Otherwise, the stock will absorb moisture from the concrete—which will announce itself later when the joints open up in the dryness of the house.

PERSONAL SAFETY EQUIPMENT

Always use appropriate safety gear to protect your eyes, your ears, and your lungs: Goggles or safety glasses, earmuff-style hearing protectors or foam plugs, and a dust mask.

Protect your knees when working near the floor to install baseboard or wainscot.

SELECTING TRIM & MATERIALS

Today we enjoy a larger selection of trim material to choose from than ever before. Man-made materials such as MDF and various urethane plastics offer many profiles and styles with virtually none of the defects that inevitably show up in traditional wood moldings. Even so, lumber millwork remains much in demand and is the only choice for naturally finished trim. The Arts and Crafts interior on the opposite page is magnificently trimmed in natural oak. Unfortunately, natural materials are steadily increasing in price as quality logs become harder for the mills to acquire. For that reason, you need to be savvy about how to select materials to reduce waste and get a good finish. Along with the actual trim, glues and fasteners are necessary for installation.

Before deciding on the materials to use, you'll also need to choose the style and scale of the trim you'll be installing. The trim style you choose should complement or harmonize with the architecture of the house. And if you're looking for a custom look, there are nearly an infinite variety of ways to use and combine trim.

CHOOSING A STYLE

If you are remodeling or adding on to an existing house that was built in a distinct architectural era, it's likely the decision about style is already made for you: you will simply match the new trim with the old. If you are not constrained to a particular style period, however, the design options are wide open. To narrow down your choices, first decide whether you are looking for a formal or informal look. Formal trim tends toward ornate, complex profiles and is often painted to put the emphasis on its form rather than the material itself. Informal trim, on the other hand, tends toward simple profiles and is often finished clear so that the grain of the wood is visible.

To develop a sense of the kind of trim that most appeals to you, leaf through home magazines or, better yet, tour finished homes. Save the magazine clippings and any snapshots you take in a scrapbook, adding notes to the images that indicate what you like (or even don't like). If your tastes run toward the contemporary you will likely find yourself drawn to flat, minimally molded trim—and not a lot of it. A good example of contemporary style trim is the wood sill of a window opening that is wrapped with sheetrock on its sides and top (which eliminates the need for any additional trimwork). The sill might be painted for an even more contemporary look.

Scaling trim

Once you have answered the question about the style of the trim, the next consideration is how to size it properly. The size of trim (or, to be precise, how its size is perceived) is directly related to the scale of its surroundings. For example, a certain size crown molding might seem too large if used in a room with 8-ft. ceilings, yet in another room with a high ceiling it might appear too small. This scaling effect is true with all trim elements: The size of the space dictates the size of the components.

As you are looking at trim examples, also be aware of how trim can affect the sense of proportion of doors, windows, walls, and even the entire room. For instance, tiny windows set in a large expanse of wall usually look "funny" and out-of-place if trimmed out with moldings sized to the windows rather than to the scale of the room. In this case, larger-than-usual moldings can help make the windows look larger and therefore in more pleasing proportion to the room. As an example of how moldings can dramatically affect the overall feeling of a room, notice how tall, monolithic-like walls are tamed to feel comfortably human-scale with the addition of chair rails, wainscoting or perhaps horizontal band boards that run over the top of the windows and doors.

ARTS AND CRAFTS

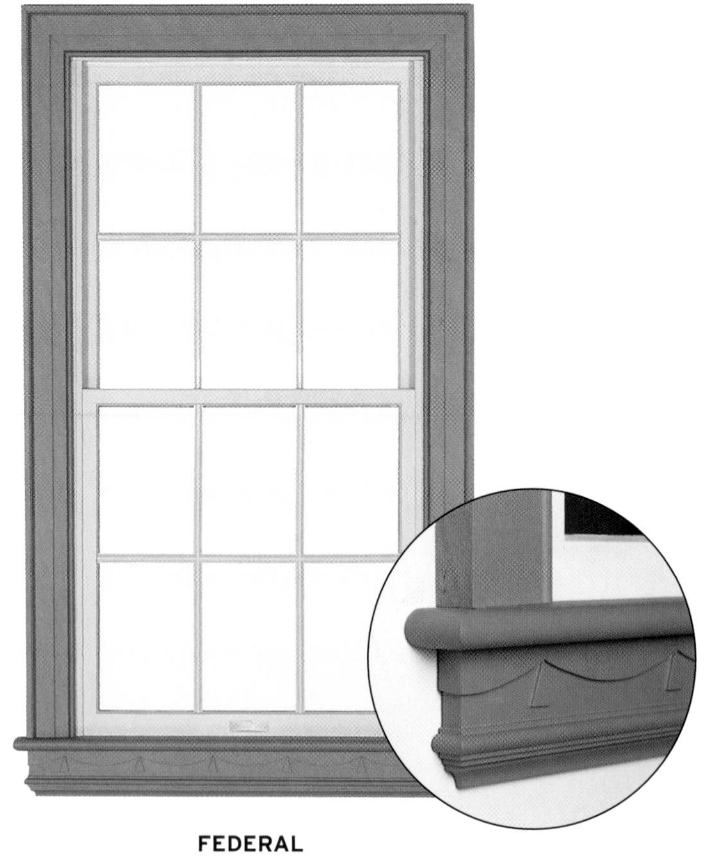

FEDERAL

VICTORIAN

CRAFTSMAN

COLONIAL

MOLDINGS

You will find that most lumberyards and building centers carry a wide selection of the moldings that are normally being used in any particular region. You should, however, be sure to check that you can obtain the moldings you want before getting too far into the design and planning process. Sometimes the material will have to be ordered or specially made, which will add to your project's time and cost. Moldings are priced by the lineal foot—the more complex and the larger in size the molding profile, the higher the cost per foot.

Router bits can shape simple profiles or be combined to make an infinite number of custom moldings.

While many suppliers typically stock moldings in a few different species of wood (most commonly hemlock and fir and finger jointed primed pine), they also will likely carry moldings in other materials besides wood. A large selection of typical molding profiles are now produced in MDF, urethane plastic, and other synthetic materials. Normally a supplier will only stock the millwork from one particular manufacturer and most likely will not have the complete line in stock but can often order what you need.

If you need something out of the ordinary a custom mill shop is ideal. These shops will work with you to help you decide exactly what you want and can mill the stock in the species and the exact profile you need. This is obviously a more costly way to go, but it may be your only option if the profile you want is not standard or if you are intent on duplicating historical moldings. However, if you own a router and a router table you can often combine basic shapes to make more complex moldings. With ingenuity, you can even duplicate historical profiles.

See the section on "Custom Moldings," pp. 208–217

What if you want the molding to go around a curved wall? Bendable moldings made of a type of urethane plastic are now available that simplify installation over or around curved surfaces. They are available in almost any molding style.

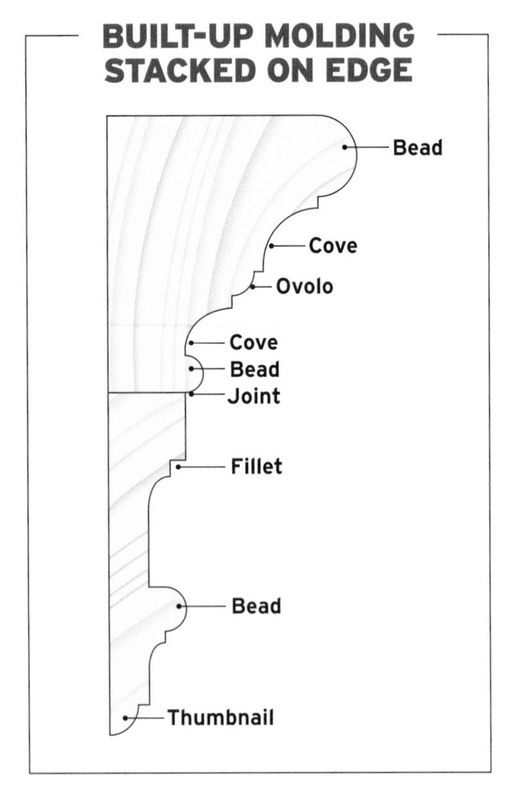

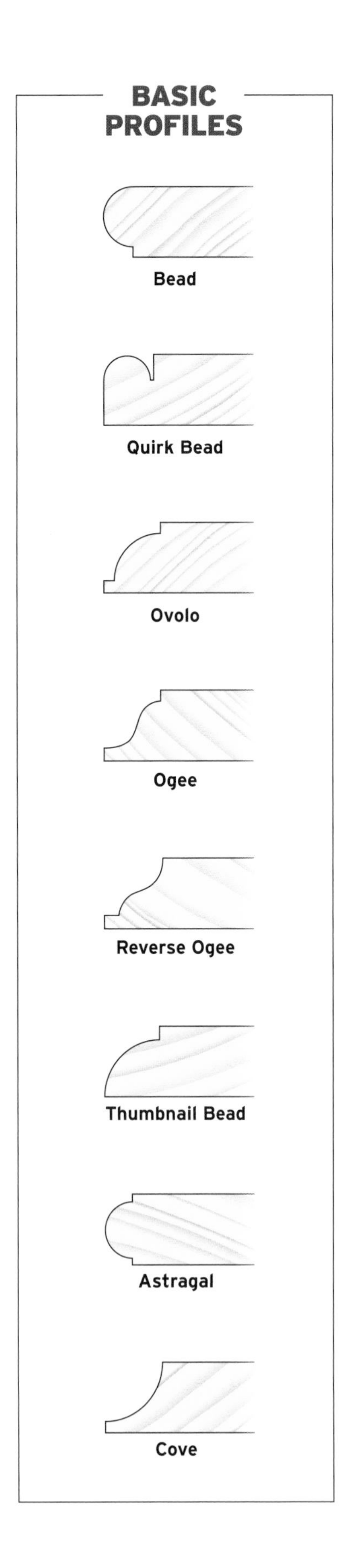

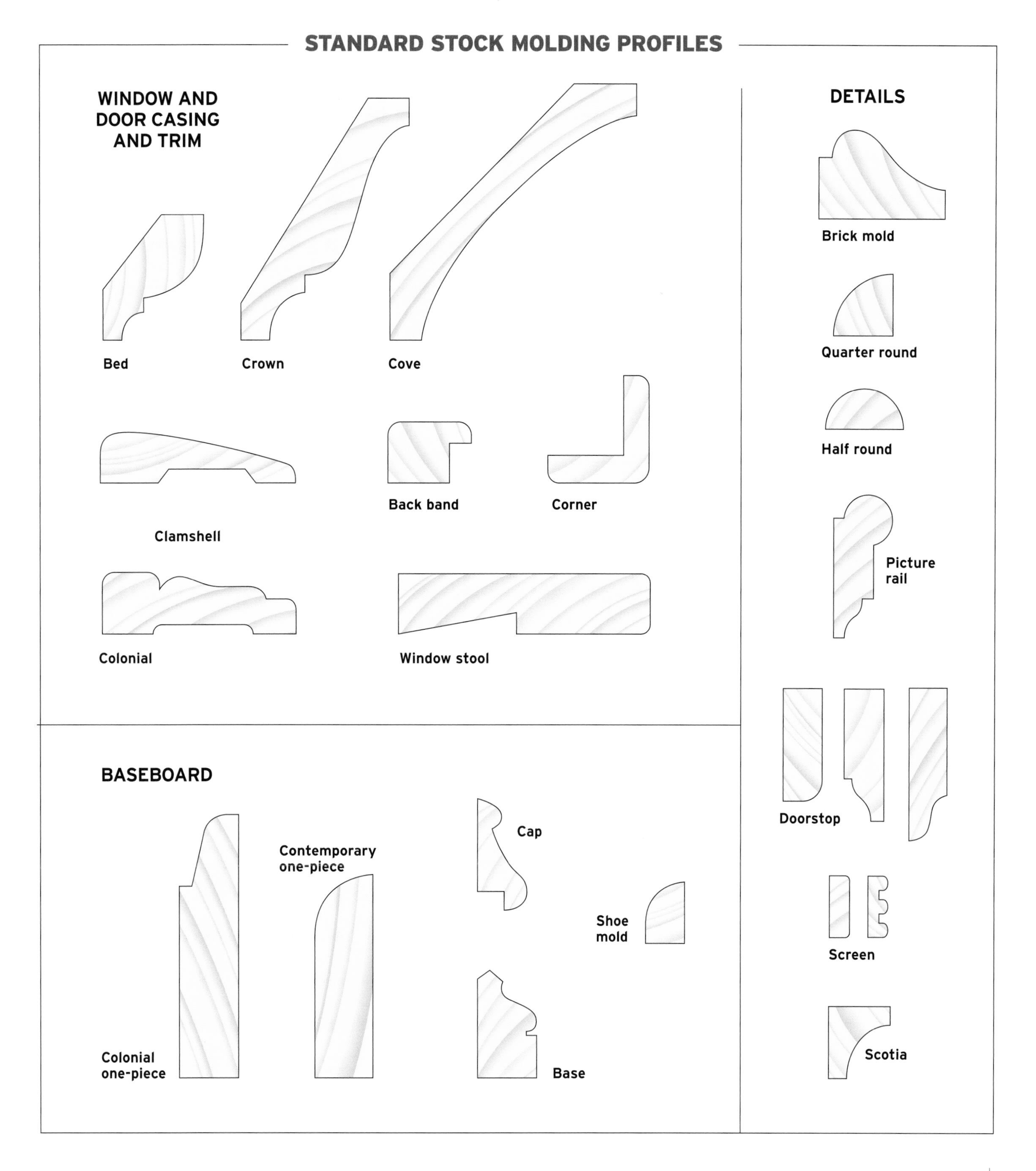
STANDARD STOCK MOLDING PROFILES
WINDOW AND DOOR CASING AND TRIM
Bed
Crown
Cove
Clamshell
Back band
Corner
Colonial
Window stool
BASEBOARD
Colonial one-piece
Contemporary one-piece
Cap
Shoe mold
Base
DETAILS
Brick mold
Quarter round
Half round
Picture rail
Doorstop
Screen
Scotia

LUMBER AND SHEET GOODS

If you choose to paint your trim, it is cheaper to use manufactured substitutes than solid wood. Keep in mind that wood is stronger and more rigid (making it a better choice for window jambs and stools) and its dust and fumes during milling processes are less toxic. If you do want to use solid wood for painting choose stable and relatively inexpensive woods such as poplar or pine. After the piece has been molded it is then preprimed. This produces a molding that is straight and free of defects—though the glue joints may transmit through the paint requiring some filling and sanding.

MDF is also a good choice for paint-grade millwork. It paints and machines well and it has a hard surface that resists dings and dents. It's also less expensive than solid wood. Its disadvantages are its weight, lack of stiffness, and susceptibility to moisture. If MDF gets wet, it expands, never to return to its original shape. I don't recommend it for window extension jambs or for stools. MDF also offers poor purchase for screws along its edges.

Lumber

When working with natural materials, it is important to be able to identify lumber defects since they affect both appearance and workability. Ideally, material used for interior trim work should be flat and straight and even-grained.

Wood, as an organic material, can change shape and size depending upon the amount of moisture it is holding relative to the surrounding air. Distortion of a board can make wood difficult—and sometimes potentially dangerous—to work with. The best strategy is not to use these boards. There are several types of distortion.

- Bow is a length-wise bend.
- Cup occurs when the edges of the board turn up.
- Twist make the board higher and lower across the diagonal.
- Crook (also called wain) is a bow along the edge, often caused by sapwood on one edge.

Sight down the end of a board to see if it's straight.

Cupping is more common in wider boards. To prevent cupping after installation on pieces of lumber 6 in. or more in width (if they will have a non-showing "back" side) rip at least three ¼-in. deep kerfs the full length of the board on the back side. This helps prevent the cupping that occurs from uneven drying, which can start as soon as the board is installed and the first coat of sealer goes onto the finished surface.

Sheet goods

The most common plywood is a 4-ft. by 8-ft. panel composed of odd-numbered layers of veneer laid up with grain running at right angles to one another (to ensure dimensional stability). Other varieties are made of single facing veneers laid over a substrate of solid wood "lumber core" or a "solid core" of MDF or particleboard. As a substrate, MDF offers the most stable and flattest surface for the facing veneers.

Cabinet grade plywood such as used for ceilings or wainscot is often faced with hardwood veneer such as oak, birch, or luan mahogany. Other species are available through special order. Plan to cover edges that will be exposed with veneer edge-banding or solid-wood strips.

MDF panels are an engineered product made entirely from waste wood—usually pine lumber ground to a fine dust and compacted with a resin bonding agent. The product is defect-free and features a very uniform density that lends itself well to shaping and cutting. Unfortunately, MDF has little structural strength and must be supported (especially when used for shelving).

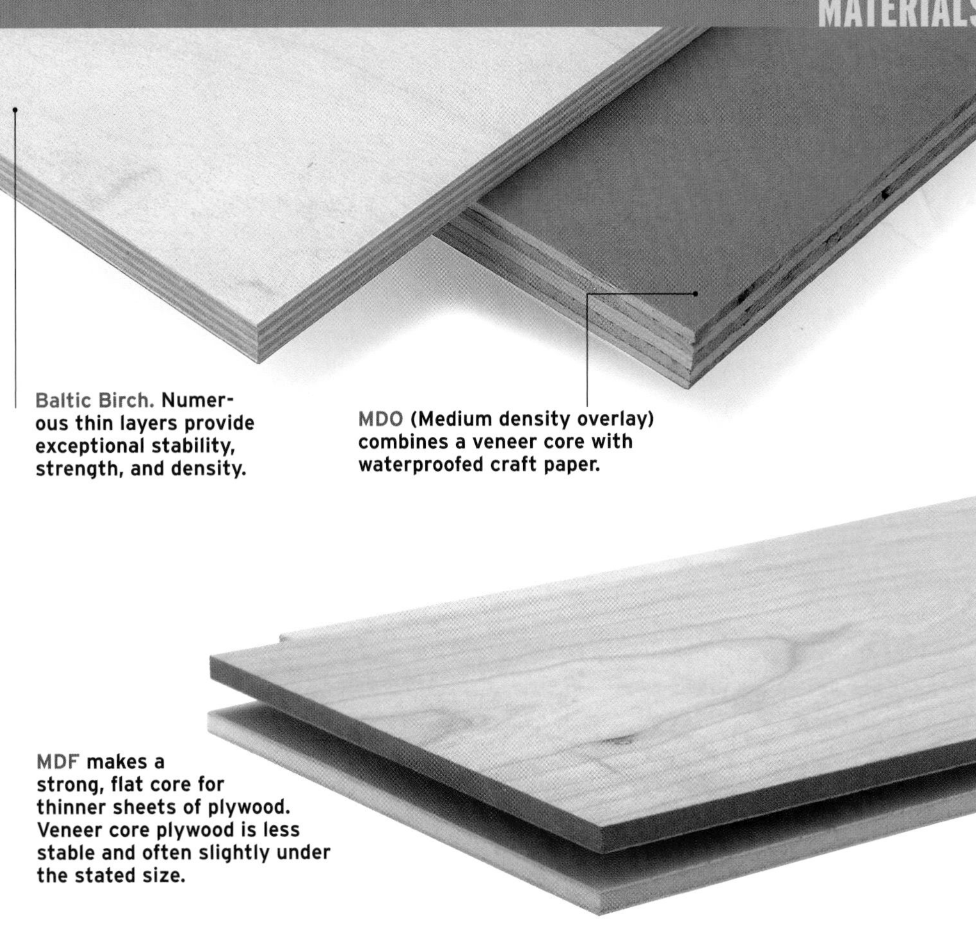

Baltic Birch. Numerous thin layers provide exceptional stability, strength, and density.

MDO (Medium density overlay) combines a veneer core with waterproofed craft paper.

MDF makes a strong, flat core for thinner sheets of plywood. Veneer core plywood is less stable and often slightly under the stated size.

Controlling Moisture Content Changes in moisture content can cause the joints in your finely fit trim job to open up. While you can't control seasonal changes in moisture, you can make sure the stock you use is as dry as possible. Protect it from rain during transit and avoid direct contact with concrete floors during storage.

Crook

WARNING A 4 ft. X 8 in. sheet of 3/4-in. thick MDF weighs about one hundred pounds! Be careful handling it. Sawing or routing it creates mountains of very fine dust, so be sure to wear a dust mask.

SYNTHETIC TRIM

Trim elements made of high-tech plastics were first developed for outdoor use. Now there are a wide variety of trim elements for paint grade indoor trim, from the simple to the incredibly ornate. Most synthetic trim is manufactured from high density polyurethane, which is rigid but lightweight. You can't cope plastic molding the way you can wood or even MDF, so inside corners of running moldings are mitered. To get a tight fitting miter, use a protractor or sliding bevel gauge to transfer the angle directly to your miter saw. Miters are glued, and many manufacturers suggest using a proprietary adhesive, so check the specifications.

Another manufactured material is "Compo," short for "composite architectural molding." Compo has been used for centuries when a "carved" element can be painted. The material comes with glue already applied on the back. It's steamed over a pan of water on a frame covered with muslin. The steam makes the element flexible and activates the glue.

Synthetic crown molding is lightweight and comes in a wide variety of profiles from the simple to the ornate.

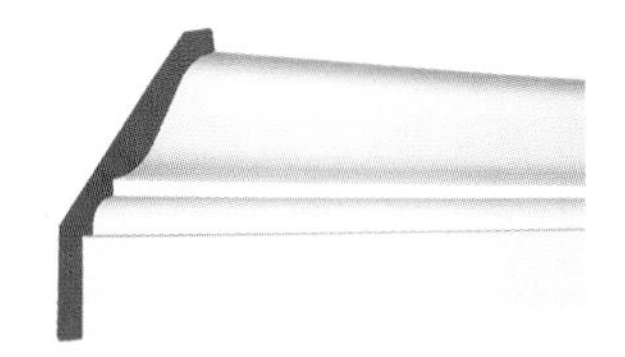

After steaming, gently apply the compo to the wood surface. A few moments of pressure bonds it in place.

GLUES

Water-based yellow glue (also known as PVA glue) is the best all-around glue for joinery. It's also the best for biscuit joinery because the water swells the compressed biscuit and locks the joint.

In trim work, glues are used to secure joints, especially to keep miters tight. While there are many types of glue available, four types are especially useful.

Aliphatic resin ("yellow") glues have become the most commonly used woodworking glue. As a rule they are stronger and set up faster than the white glues. However, white glues set up more slowly than yellow glues, allowing a little more flexibility and a more relaxed pace when gluing up large or complicated assemblies.

Cyanoacrylate ("super" or "instant") glues are a lifesaver on the job for such things as gluing edge splinters and splits in the wood. They are strong, have reasonably good gap filling ability and when used with an accelerator set almost instantly. Read the precautionary warnings and wear gloves and eye protection when using these glues.

Polyurethane glues are a relatively new product. I have not found them to be any better than yellow glue for interior carpentry. Polyurethane glues have some gap filling qualities, owing to its tendency to foam when exposed to air and moisture. These are harder to clean off the workpieces and your hands.

Construction adhesive (typically available in caulking tubelike applicators) is a great product. Though it takes a while to set up, it's great for filling in behind a cavity or gluing material to an irregular surface. Once dried it is extremely strong. Most of the brands are petroleum based, however, so avoid the fumes.

TRADE SECRET When using natural finishes you must remove all the glue from the surface or you will see it under the finish. Sometimes it's easier to let the glue dry and then scrape it off, instead of smearing it over a wider surface with a wet rag.

GLUING MITERS

1 Apply glue to each face of the miter and spread it evenly.

2 Pinch clamps help hold the miter tight.

3 Use headless pin nails to reinforce the joint, nailing across the miter.

FASTENERS

The types of fasteners typically used to install trim include nails, staples, and screws. Traditional hammer-driven finish nails are categorized by "penny" size—the larger the penny number the larger the nail. Pneumatic-driven nails are identified by gauge and length. The larger the gauge the smaller the diameter of the nail will be. Pneumatically driven staples work well for attaching plywood; however, due to the large hole created in the face of the material they are suitable only where they will not be visible.

While screws are not very practical for installing all types of moldings they can offer certain advantages over nails with their ability to draw large moldings together for a tight fit. Installing and assembling moldings with screws also allows you to tweak the moldings to fit if necessary or to remove them with ease entirely. And using screws on a glued up assembly often eliminates the need to use clamps. The screws I use for trim are either bugle head or common head and are chosen by the following characteristics:

Gauge refers to the overall diameter of the screw, the larger the gauge number the larger the diameter and strength. The larger gauges are also available in longer lengths. Number 8 or number 6 screws are a good size for most trim and general assembly work.

The length of a screw for a particular job can be determined by this rule of thumb: Two thirds of the screw's length should protrude into the surface being fastened to.

Drive types useful for trim work are the Phillips and the square drive. I don't even bother with the slotted drive, as they are difficult to use with a power driver. While both Phillips and square drive work well, the latter more firmly engages the driver bit, reducing slippage and stripping. The size of the driver bit varies with the screw gauge from #0 to #4 with #2 being the most common.

Screw heads commonly used for installing trim are the flat, pan, and trim head. Flat heads are used for general fastening. The head is either kept flush to the surface of the wood or countersunk and plugged. Pan heads are flat on the bottom surface of the head and are generally used for attaching hardware. They are also used almost exclusively for pocket-screw joinery. Pocket hole screws are sometimes available in fillister head or washer head styles. The washer prevents the screw from penetrating too deeply into soft woods. Trim head screws have a very small head much like a traditional finish nail. They are driven below the surface of the wood leaving a hole to be filled. There is a new version of these designed for installing composite decking that I have had great success with for fine adjustments of trim.

See "Double Threaded Screw," p. 184.

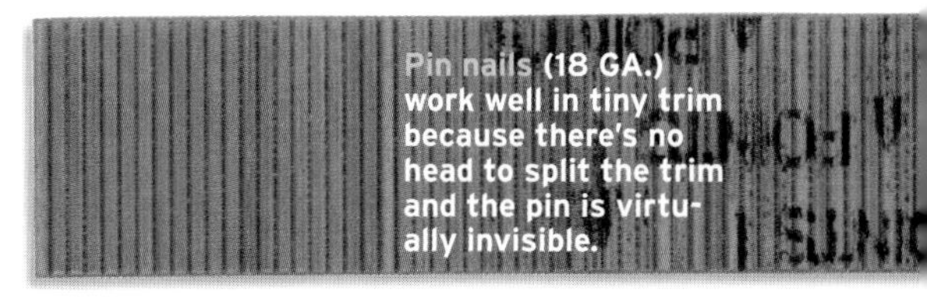

Pin nails (18 GA.) work well in tiny trim because there's no head to split the trim and the pin is virtually invisible.

Slight-head pins (18 GA.) resist pull through better than headless pins.

Brad nails (18 GA.) work for low-stress trim. Brads come in lengths from 5/8 in. to 1 9/16 in.

Finish nails (16 GA.) are the best for all around trim work. Nail lengths from 1 in. to 2 1/2 in.

Finish nails (15 GA.) offer structural stength for jobs like installing solid core doors. Nail lengths from 1 1/4 to 2 1/2 in.

Flat (bugel) head

Finish head

Pan head

Washer head

Filister head

TOOLS & TECHNIQUES

A SMALL TRIM JOB, LIKE REPLACING casing around a single window or door, can be accomplished with just a few simple tools: a measuring tape, a miter box and backsaw, a hammer, and some nail sets. But once you get beyond the basics, it helps to have the right tools.

Most of the tools used for installing trim are also used for general carpentry and home repair, so you'll use them for other jobs. Invest in quality tools and they will last you a lifetime. But how do you judge quality?

Surprisingly, the most expensive tools aren't always the best value. Often you are paying for extra features you may never need. Evaluate your future projects and balance them against your investment. You can't go wrong if you stick to trusted brands and purchase mid-price tools.

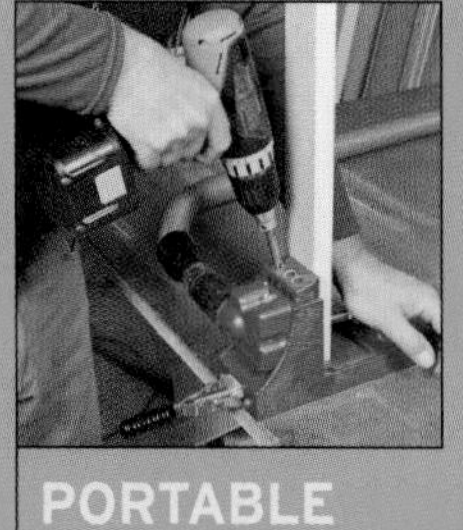

TRIM CARPENTRY

POWER MITER SAW

POWER MITER SAWS EXCEL AT MAKING ACCURATE CROSSCUTS AND CUTTING MITERS. A SLIDING COMPOUND MITER SAW SUCH AS SHOWN HERE CAN ALSO CUT COMPOUND ANGLES FOR CROWN MOLDING.

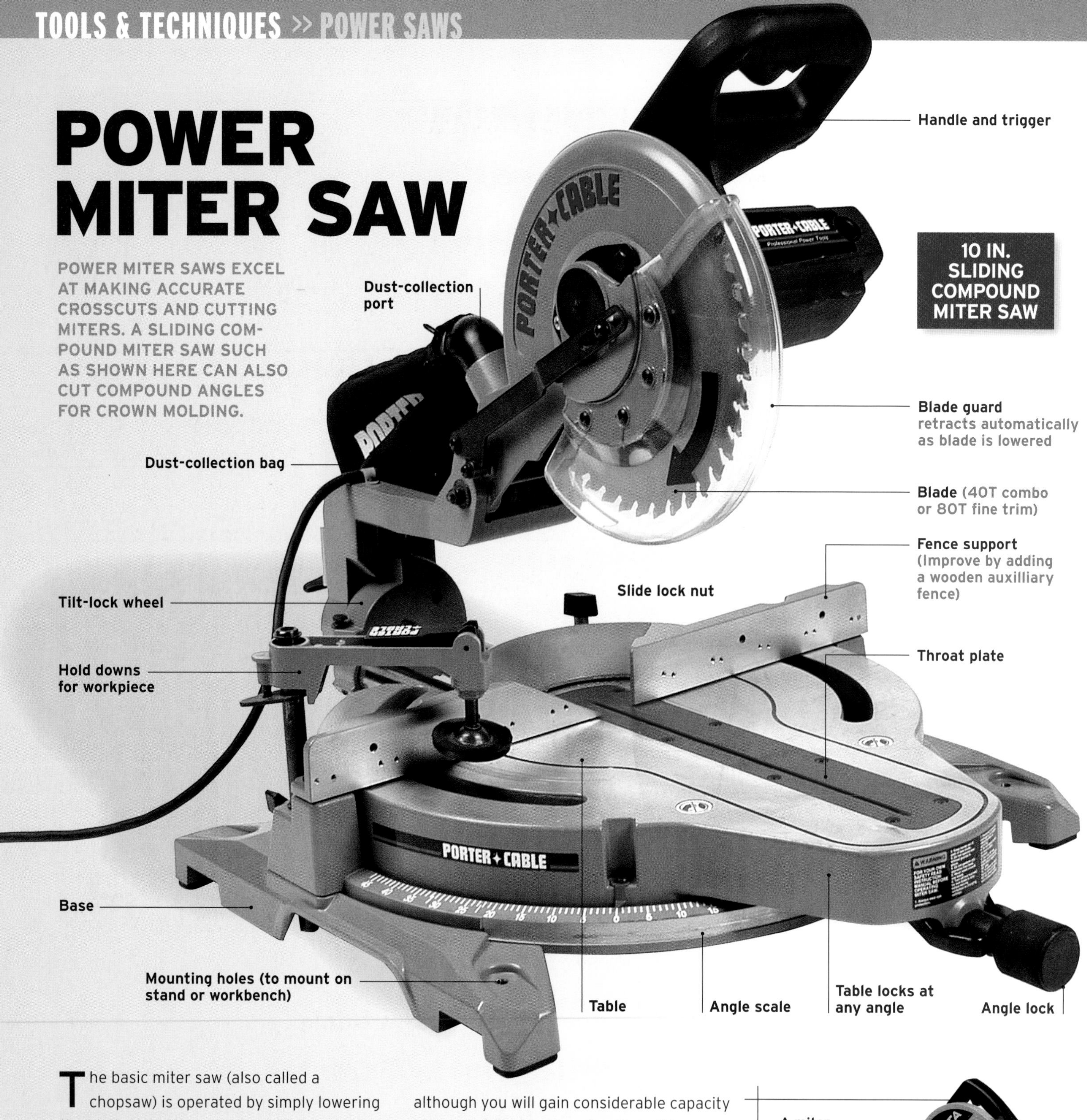

The basic miter saw (also called a chopsaw) is operated by simply lowering the blade onto the workpiece. The maximum width of cut is determined by its blade size. A sliding compound miter saw enables the saw to slide out over the bed and also allows the saw blade to be tilted in one direction so it can cut two angles simultaneously. The dual compound version features a blade that can be tilted in both directions. For basic trim installation a chopsaw is adequate, although you will gain considerable capacity and versatility by upgrading to a slide saw. The size of the blade significantly affects both the price and the physical size of the saw. For most trim carpentry a 7½ in. saw will do the job, but a 10 in. saw is capable of cutting thicker materials and larger moldings in an upright position. Look for a saw that has good visibility of the cut line through the guard.

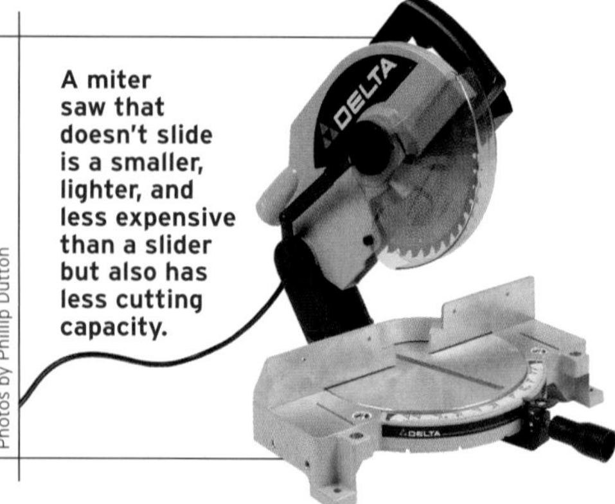

A miter saw that doesn't slide is a smaller, lighter, and less expensive than a slider but also has less cutting capacity.

Photos by Phillip Dutton

CROSSCUTTING WITH A MITER SAW

To set up the cut, mark the board for length and line the mark up with the edge of the saw blade making sure that you are cutting on the correct side of the line ❶. Hold the material tight against the fence with your left hand to the left side of the blade and operate the saw with your right. Never cross your arms to hold the material as this puts them in the potential path of the saw blade. Allow the blade to reach full speed before starting the cut. Once the cut is complete release the trigger. Be sure the saw blade has come to a complete stop before raising it up out of the workpiece ❷.

1 Hold material tight against the fence and align saw blade with line.

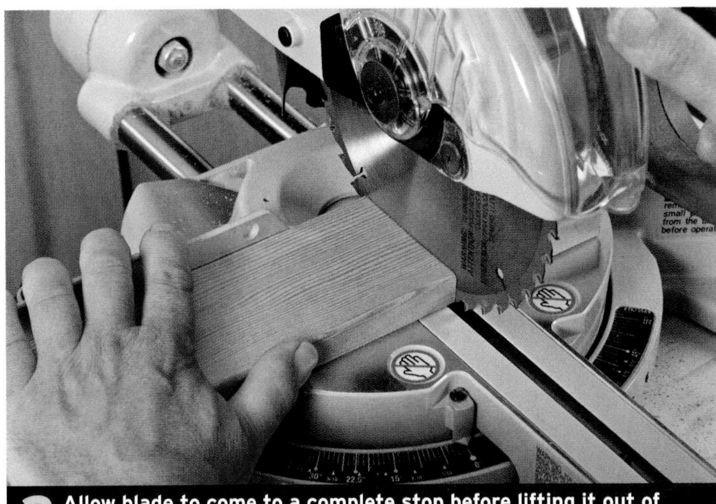

2 Allow blade to come to a complete stop before lifting it out of the works.

INVEST IN A MITER SAW STAND

As a professional there were a few things that totally changed the way I worked. One of those was a dedicated miter saw stand and extension wings with a repetitive cut feature. If you are going to be using a miter saw for anything more than an occasional small project, it's a great investment. A proper stand promotes the safe use of these saws by eliminating the difficulty of supporting long workpieces. It also ensures absolute repeatability of multiple piece cuts and dramatically increases the speed at which you can work. If the extension wings are solid rather than simply a bar to hold the stock up they can also be used as a bench while coping baseboard or other tasks. With a little ingenuity, a homemade version can be made in lieu of a commercial version.

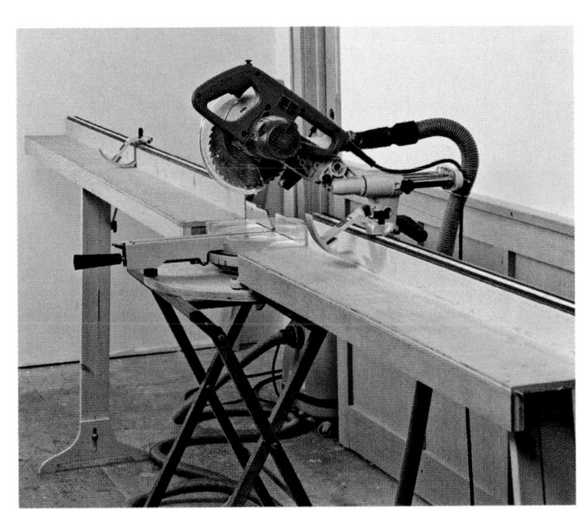

This $7^1/_2$ in. saw is adequate for most trim work.

Miter saw safety

- Position your hands so you don't cross your arms.
- Hold the workpiece securely.
- Keep your hand at least 3 in. from the blade.
- Allow the blade to come to a complete stop before lifting it.

CUTTING MITERS

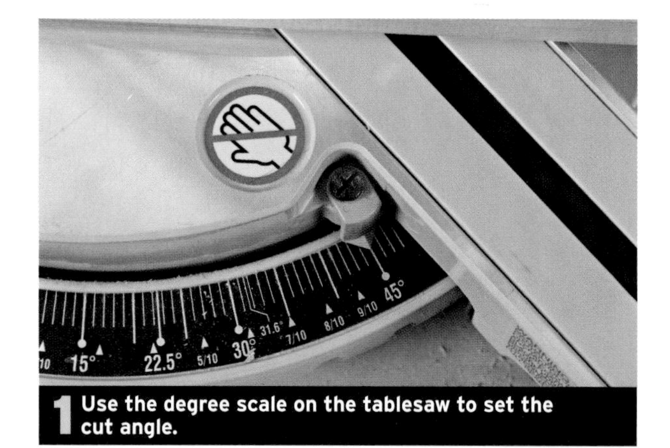

1 Use the degree scale on the tablesaw to set the cut angle.

To cut a miter angle on a miter saw start by loosening the table and rotate it until the saw pointer is aligned with the desired degree setting on the table scale ❶. Now lock the table in position. Hold the stock tight to the fence with your left hand at a safe distance from the path of the blade (a good rule of thumb is the width of your hand), then start the saw. Allow it to come to full speed ❷. Using a steady downward pressure, lower the blade into the material to make the cut. Again, make sure to let the blade come to a full stop before returning the saw to its resting position ❸.

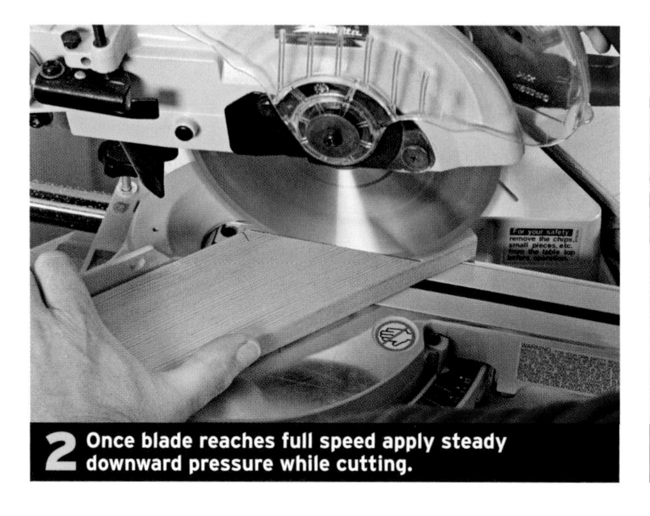
2 Once blade reaches full speed apply steady downward pressure while cutting.

3 Release switch and allow blade to stop before lifting from material.

CUTTING WIDE STOCK

A slide miter saw gives you the ability to cut wider material. Once again make sure your hands and arms are away from the path of the blade as you hold the stock tight against the fence. Allow the blade to reach full speed before starting the cut at the edge of the board closest to you. Now lower the saw into the material and slowly push the blade back toward the fence ❶. This technique eliminates any tendency for the saw to climb and pull itself toward you while cutting. Once again allow the blade to stop fully before raising it from the workpiece ❷.

1 With blade at full speed start cut at edge closest to operator.

2 Push saw slowly through the cut allowing the blade to stop before lifting.

CUTTING COMPOUND MITERS

A A standard miter saw **requires that you tilt the molding at an angle to the fence and the saw table.**

1 On a compound miter saw one option is to cut the material flat on the table. First set the miter angle.

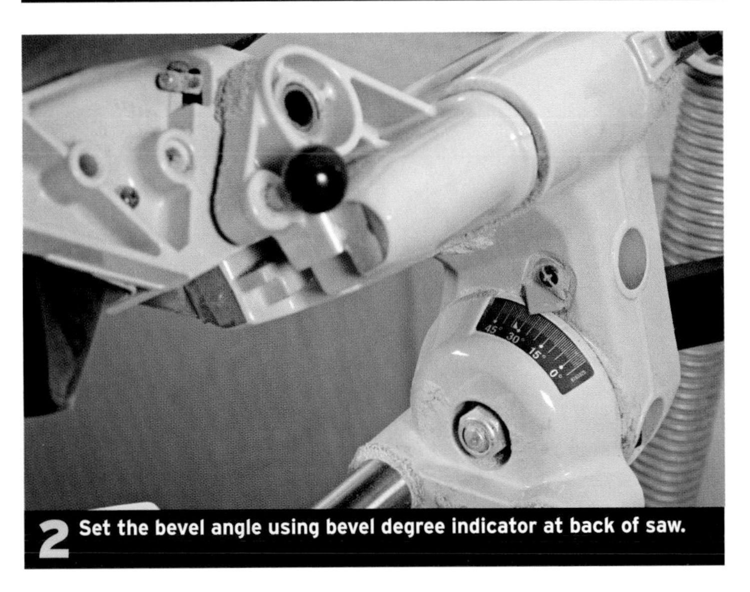

2 Set the bevel angle using bevel degree indicator at back of saw.

A compound miter occurs when there is an angle across the face of the molding (the miter angle) and an angle through the thickness of the stock (the bevel angle). To cut a compound miter on a standard miter saw you can incline the molding at the desired angle between the fence and the saw table with the saw set to the correct miter angle. The resulting cut will be angled in two planes **A**.

➔ See "Cutting Crown in Position on a Miter Saw," p. 132.

On a slide miter saw you can cut a compound miter as described above or, when working with wider moldings, by laying them flat on the saw table.

➔ See "Determining and Transferring Angles," p. 133.

To make this type of cut, set both the saw table's miter angle **1** and the saw bevel angle **2** to the appropriate angles.

ADDING AN AUXILIARY FENCE

The standard fence on the saw usually has holes so that you can attach a wooden fence with screws. The advantages of an auxiliary fence are many. It adds extra height to better support thicker stock or moldings. In the case of cutting compound miters, you can draw a line on the fence so that you can consistently align the molding at the proper angle.

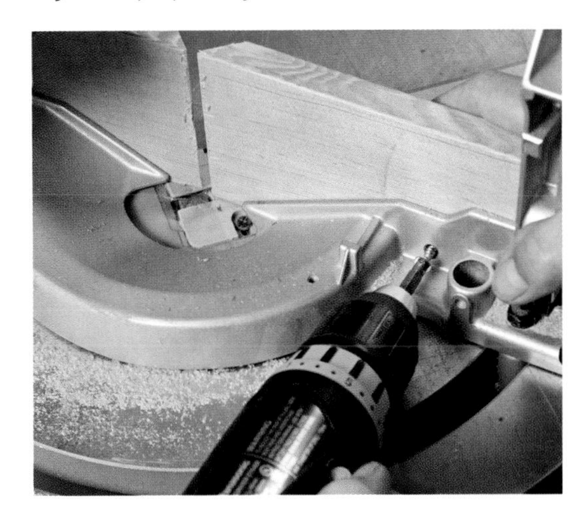

USING THE DEPTH STOP ON A SLIDE MITER SAW

Slide miter saws feature a depth stop that can be used to cut partially through a board—a typical application would be creating an end rabbet. Start by marking the depth of the cut on the end of either a piece of scrap or the stock to be cut. Then set the board on the table and lower the blade until the teeth are even with the mark ❶. Holding the saw in this position, adjust the adjustment screw on the stop until it stops the saw at the correct depth ❷. When the saw is set to cut anything less than full depth, the curvature of the blade makes a cut that rises a bit immediately next to the fence; it is necessary to insert a spacer between the material and the saw fence to allow the saw to cut an even depth across the full width of the board ❸.

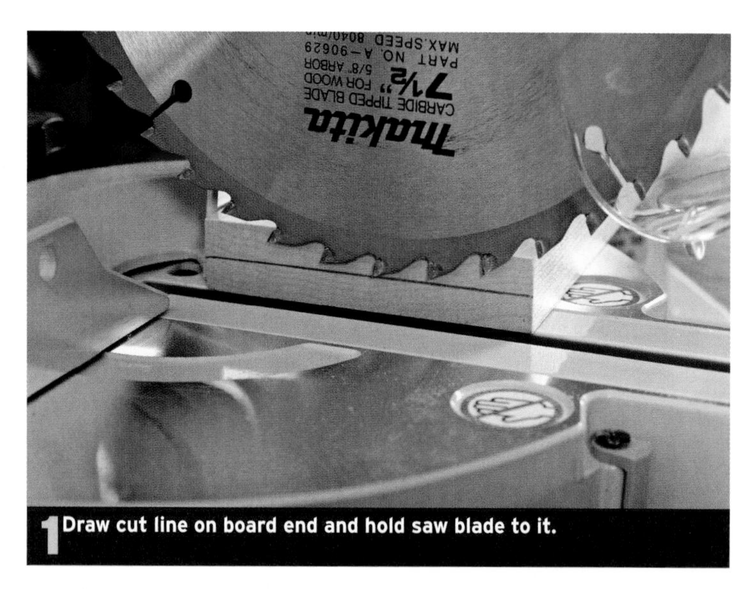

1 Draw cut line on board end and hold saw blade to it.

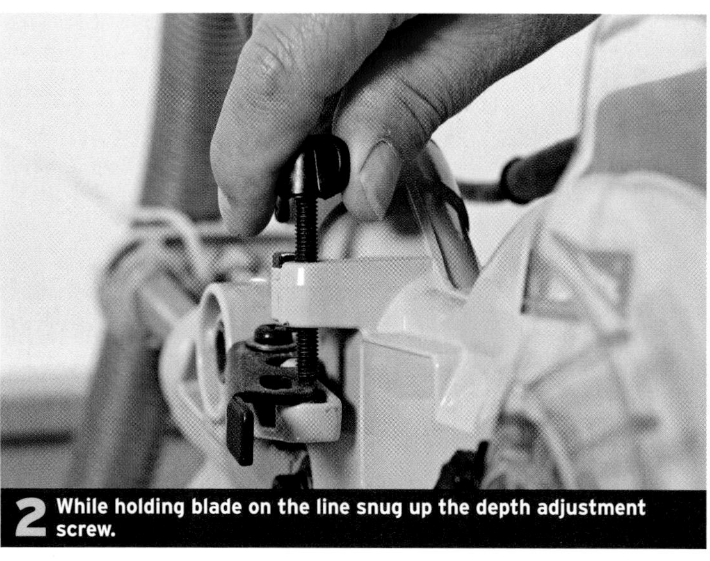

2 While holding blade on the line snug up the depth adjustment screw.

3 Using a spacer between fence and material enables saw to complete cut.

JOINTS THAT CAN BE CUT WITH A DEPTH STOP

END RABBET

DADO

SHOULDERED DADO

CIRCULAR SAW

CIRCULAR SAWS, PROPERLY GUIDED, CAN MAKE ACCURATE CUTS. THE SAW SHOWN IS A SIZE AND TYPE COMMONLY USED IN GENERAL CARPENTRY, SO ACCESSORIES AND BLADES ARE EASY TO FIND.

Photo by Phillip Dutton

Buying a solid professional grade tool will stand you in good stead for years to come. A 7¼ in. circular saw is the most universal size, making it easy to find blades. For extensive trim work, buy a corded model. The sole of the saw needs to be strong, straight, and flat. Also look for a depth of cut adjustment that's easy to operate and a scale that's clear and readable. These saws come blade-right and blade-left (not always available). The standard is blade right, but blade left provides better visibility for right-handers.

While sidewinders are the most commonly used construction tool on the East Coast, West Coast builders like worm drive saws. This preference seems totally regional, but one way to choose is to ask the advice of a local carpenter or the manager of the tool department in your homecenter. Keep to nationally recognized brands, and choose a mid-price professional grade tool. You'll use your circular saw more for general construction than for trim work, and it needs to be sturdy and well-made to withstand the demands of framing and breaking down sheet goods. Fitted with special blades, it can also cut drywall and cement board for tiling.

GUIDING A CIRCULAR SAW

A A speed square **makes an excellent guide for short crosscuts.**

B A site built saw guide **is capable of producing very straight cuts.**

C This commercial saw guide **also serves as a clamp and a straightedge.**

D The factory-supplied guide **is best used when absolutely straight is not necessary.**

For accurate, straight cuts using a circular saw a guide is a necessity. When crosscutting up to 12 in. a large Speed Square® works well to guide the saw. Hold the flange of the square against the edge of the board and align the other edge of the blade on the cut mark **A**. For making wider crosscuts clamp a shop-built saw guide to the board **B** or use a commercial straight edge-clamping device **C**. Ripping lengths of material can be done in several ways. The easiest way suitable only for rough work is to free hand the cut. Most saws come with a ripping guide that is temporarily installed on the saw. It rides along the edge of the material as the cut is being made **D**. These guides produce a reasonably straight cut, but don't rely on them when precision is critical. Instead, use a full-length ripping guide clamped to the material. Any straight material will work or better yet, make a set of both short and long guides for your saw.

➔ See "Making Site-Built Crosscutting/ Ripping Guides," p. 25.

WARNING

When cutting with a circular saw it is important for safety adjust the blade depth so that it protrudes no more than 1/4 in. deeper than the material being cut. This keeps most of the blade buried in the material and allows the blade guard to operate smoothly when starting the cut.

BREAKING DOWN PLYWOOD SHEETS

Ripping large, and especially full-size, sheets of plywood on a portable table saw should be avoided if possible. It's cumbersome and dangerous. Breaking down the sheets into more manageable sizes with a circular saw is the best way to go. You'll need a pair of sawhorses with some cross supports to support the plywood ❶ and a saw guide to ensure accurate, straight cuts. The simplest guide is a straight piece of plywood of sufficient length ❷. To use such a guide, first determine the offset amount between the edge of the blade and the edge of the saw table as you must allow for that distance when setting up for the cut. With a little more work, you can make a much easier-to-use guide that is aligned directly onto the cut mark avoiding any chance for error.

Of course, you can also simply cut the piece free hand—just be sure to leave a little bit extra material so you can re-cut the piece on the table saw by orienting the straight edge of the plywood against the saw fence.

TRADE SECRET

I often use a sheet of 1-in. thick insulation foam board set on a sheet of plywood as a support, setting the saw blade to cut about 1/8-in. into the foam.

1 A support system will keep plywood pieces from falling away after cutting.

MAKING SITE-BUILT CROSSCUTTING/RIPPING GUIDES

To build a cutting guide choose a piece of 1/2 in. or 3/8 in. interior plywood that is smooth on both sides and is wide enough to hold the saw base, the guide strip and several inches on the other side of the guide strip for clamping. The guide strip should be a straight piece of wood approximately 1/2 in. thick by 1 1/2 in. wide. Screw the guide strip to the base plywood far enough from the edge so that the initial cut of the saw using the guide will cut off a portion of the base plywood leaving an edge that is perfectly in line with the edge of the saw blade.

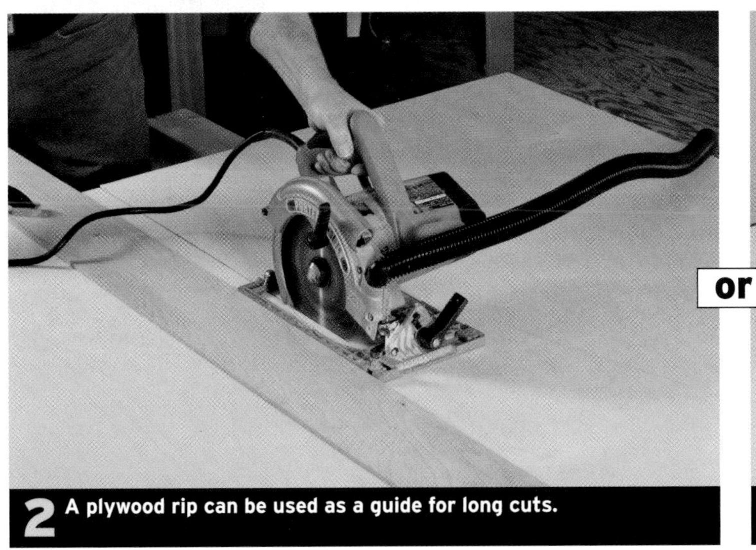

2 A plywood rip can be used as a guide for long cuts.

or

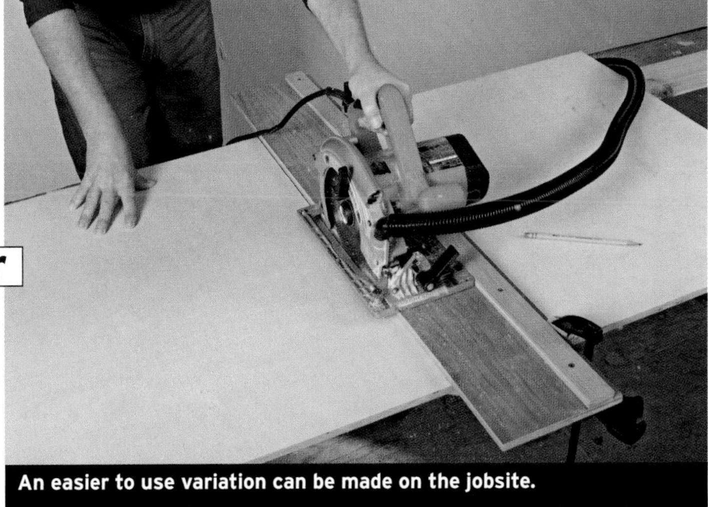

An easier to use variation can be made on the jobsite.

TABLESAW

FOR ACCURATE RIP CUTS, SUCH AS SEPARATING A SMALL PIECE OF MOLDING FROM A PIECE OF LARGER STOCK, THE TABLESAW CAN'T BE BEAT. IT WILL ALSO CROSSCUT AND MITER WHEN OUTFITTED WITH THE RIGHT JIGS.

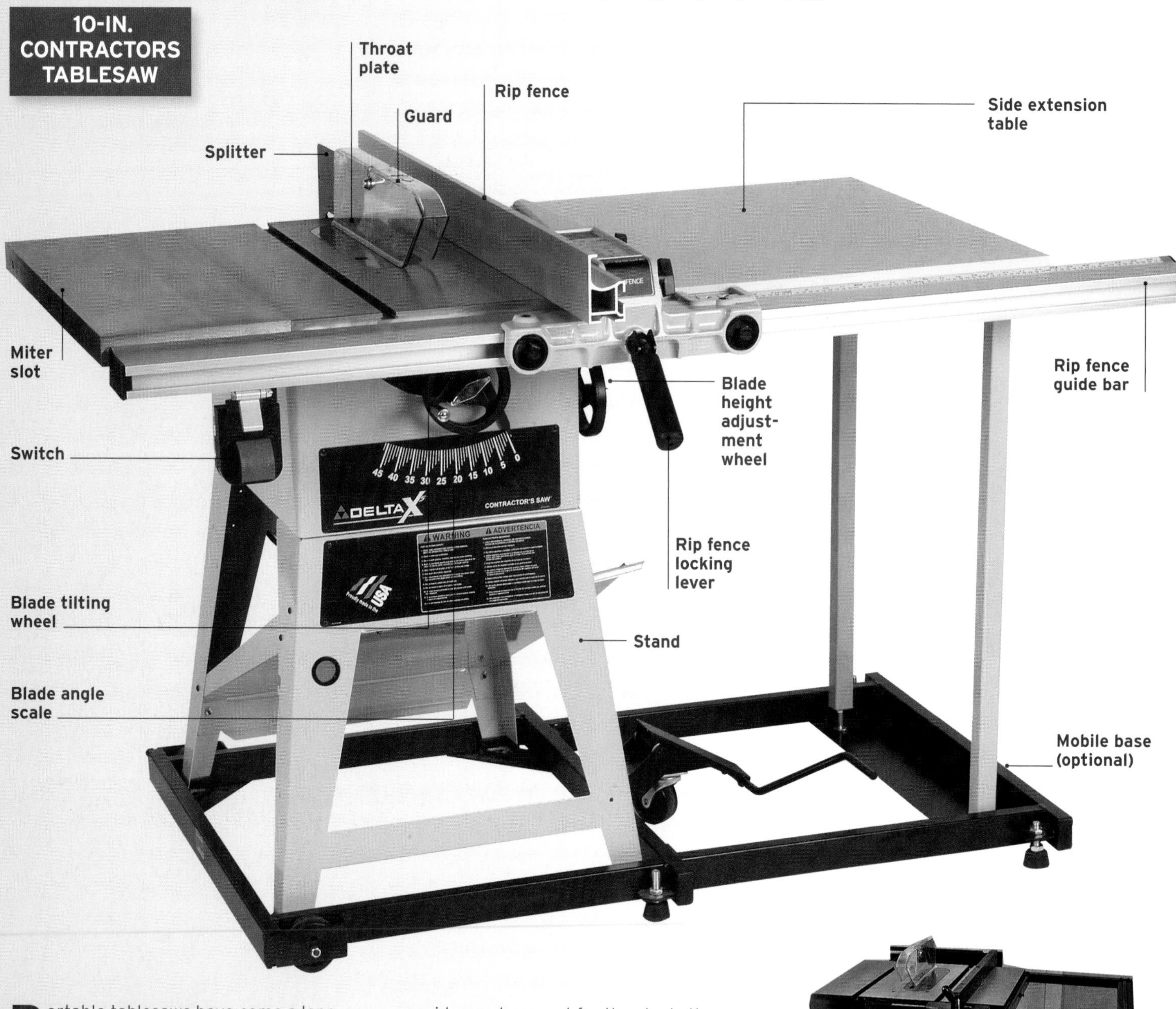

Portable tablesaws have come a long way in quality and function since I started doing carpentry. If you stick with the better-known brands you'll get a high quality, durable tool that you'll most likely never wear out. Table saws are available in blade sizes ranging from 7½ in. to 10 in.

Before purchasing a saw check out the following features: The size of the table, the fence action, and the dust management. To provide good support for the stock, the larger the table the better (although avoid making your choice on table size alone). The rip fence should move easily and it should lock into position securely and it should feature an easily read scale cursor. Finally, be sure that there is a vacuum port for dust collection so you can minimize the amount of sawdust entering your working environment.

A benchtop saw with a flat table and reliable rip fence is both portable and less expensive.

Photos by Phillip Dutton

CROSSCUTTING ON A TABLESAW

The miter gauge supplied with all saws will likely need to be fine-tuned by adjusting the stops and any slop in the guide bar in order to produce accurate crosscuts. While the miter gauge allows you to cut straight cuts and miters on small workpieces, it can't support very long stock. For that you will need to add a longer auxiliary fence to the miter gauge or switch to a miter saw.

To crosscut with the miter gauge, begin by marking where you want. Raise the blade to be approximately 1/4 in. above the workpiece. Then with the power off, position the workpiece so that blade will fall on the waste side of the line ❶. Hold the workpiece firmly against the fence of the miter with your left hand. Turn on the power, and then holding the stock firmly against the gauge, feed the stock smoothly into the blade, using your right hand to push the miter gauge forward ❷. Push the piece past the blade. Once the cut is made, use your left hand to move the stock slightly away from the blade ❸. Moving the stock away prevents splintering and double-cutting, when you pull the miter gauge back past the blade ❹.

WARNING In all photos the guard and splitter has been removed for clarity. Always use appropriate guards when using power tools.

1 Align the workpiece to your mark so that blade will cut on the waste side of the line.

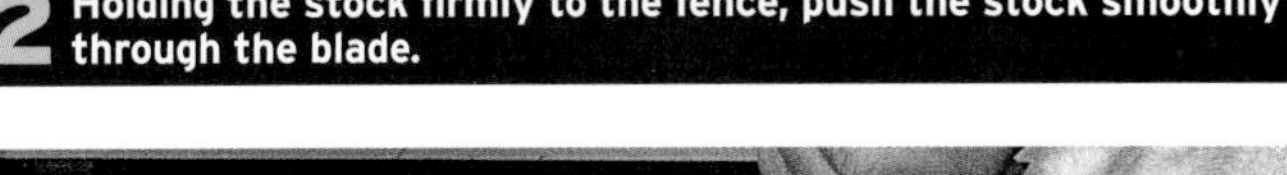

2 Holding the stock firmly to the fence, push the stock smoothly through the blade.

3 Move the dimensioned piece slightly away from the blade after the cut is finished.

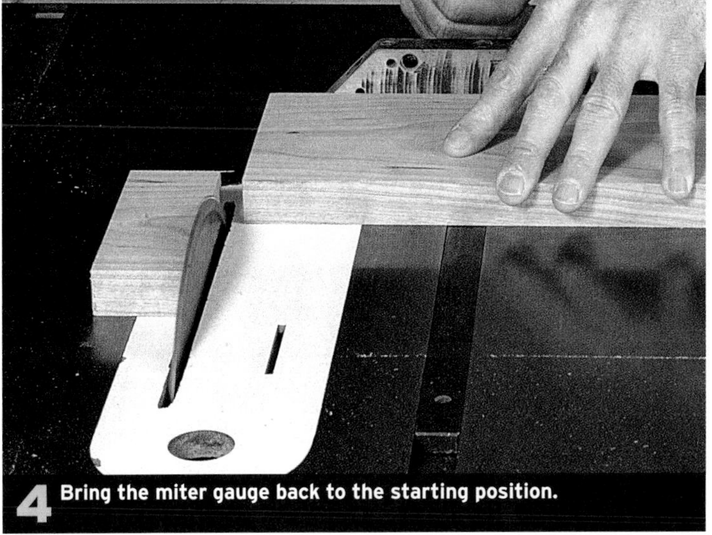

4 Bring the miter gauge back to the starting position.

Photos by Andy Rae

RIPPING WITH A TABLESAW

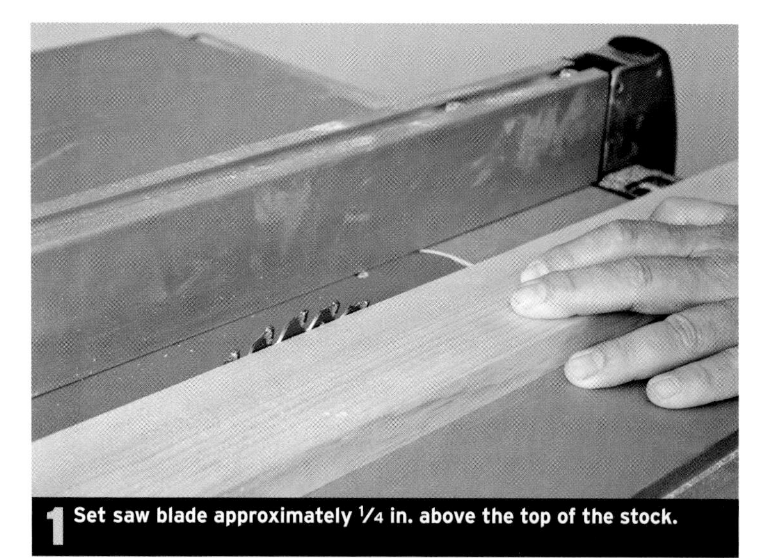

1 Set saw blade approximately 1/4 in. above the top of the stock.

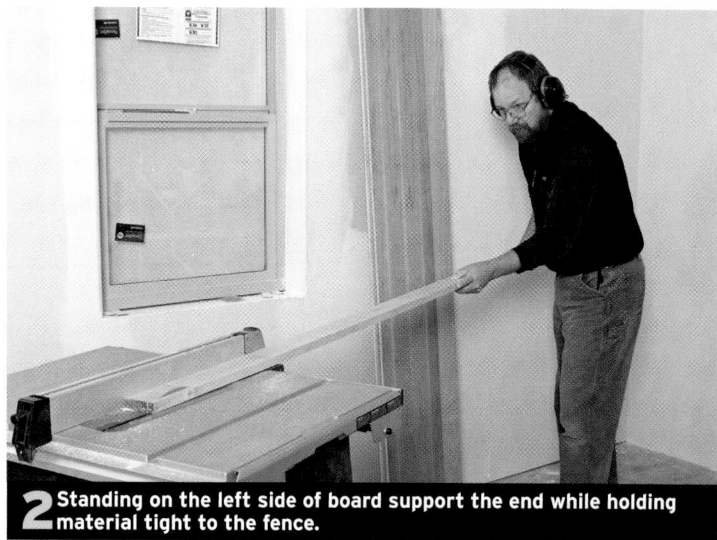

2 Standing on the left side of board support the end while holding material tight to the fence.

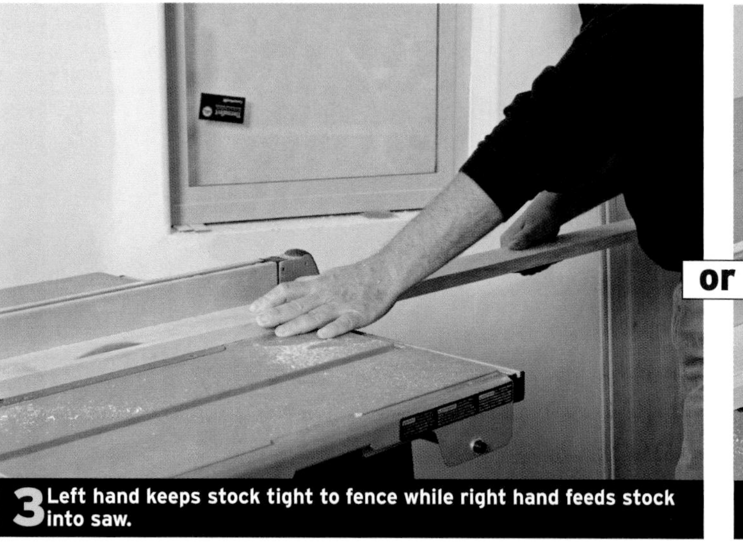

3 Left hand keeps stock tight to fence while right hand feeds stock into saw.

or

A feather board keeps the board tight to the fence and your fingers safe.

4 Remove left hand and keeping right hand clear of blade use right to finish the cut.

or

When ripping narrow stock, use a push stick to feed material as cut nears the blade.

Before ripping stock to width, be sure to make one edge straight and square. Place this edge against the fence and set the blade so that it only protrudes above the stock by about 1/4 in. ❶. Now stand to the left side of the blade and while holding the board tight to the fence feed it into the blade ❷. Push the board tight to the fence with your left hand (well away from the blade) and use your right hand to feed the stock into the saw ❸. Or use a feather board to hold the stock tightly to the fence. Once the board gets close to the end of the cut remove your left hand and continue using your right hand (again held away from the location of the blade) to push the board through the saw blade to finish the cut ❹. When ripping narrow stock use a push stick to feed the stock as it nears the blade.

SITUATIONS THAT CAN CAUSE KICKBACK

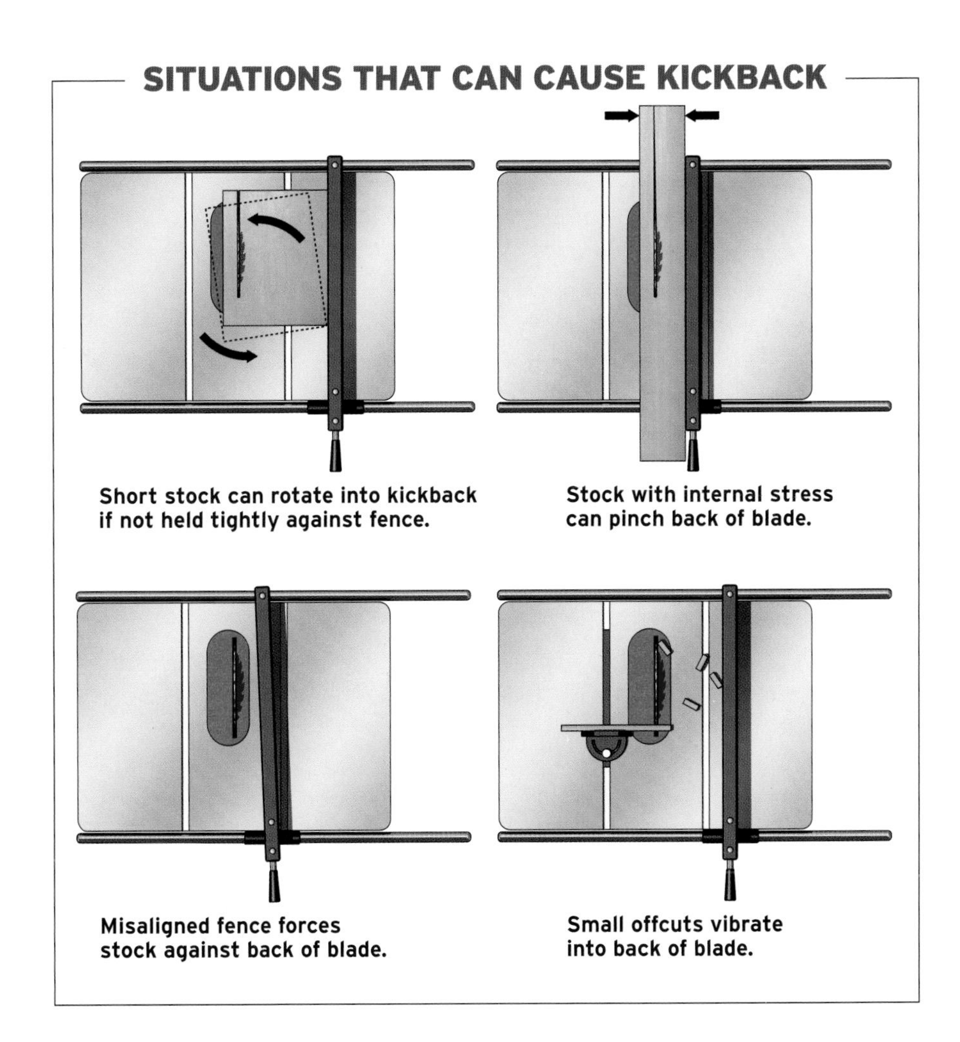

Short stock can rotate into kickback if not held tightly against fence.

Stock with internal stress can pinch back of blade.

Misaligned fence forces stock against back of blade.

Small offcuts vibrate into back of blade.

Tablesaw safety

The tablesaw is one of the more dangerous tools in carpentry. Safety guards work, but do not depend on them to prevent you from being injured. Here are some basic rules:

- **Always wear both ear and eye protection while operating the tablesaw.**
- **Use a push stick when feeding narrow material past the blade.**
- **Do not operate the saw without the guard and splitter.**
- **Do not cut with the blade protruding more than 1/4 in. higher than the material.**
- **Use a feather board to hold stock tight to fence.**
- **Never force stock through a cut.**

WARNING When ripping a long board, it's important to have outfeed support at the far end of the board. Otherwise the board can drop off, ruining the cut, or worse yet, catch on the blade and kick back at you. A permanent outfeed table is ideal, but when working on a jobsite a properly weighted and oriented roller stand can serve the purpose. Make sure to adjust the height of the outfeed support so that the stock feeds smoothly over it. Thin boards are more flexible and can dip, so plan ahead.

CUTTING EDGE RABBETS ON A TABLESAW

1 Set blade height to the marked rabbet lines for the first cut.

2 Set fence so blade lines up on the waste side of line.

Edge rabbets have a variety of uses in trim work. For example, rabbeted baseboard is used to support tongue-and-groove boards in wainscot.

➔ **See "Making Rabbeted Baseboards," p. 148.**

A beveled rabbet is also used to make jamb extensions for windows. In this case the blade is angled by 3 degrees from perpendicular.

➔ **See "Making and Installing Rabbeted Jambs," p. 62.**

Cutting edge rabbets on a tablesaw involve making two cuts, one with the board flat on the saw and the other with the board held on its edge. Start by marking out the rabbet on the end of the board and using these marks to set the blade to the correct height ❶. Now again using the marked outline, set the fence to the correct distance from the blade ❷. Make the first cut with the board laying flat on the table ❸. Then reset both the blade depth and the fence (as necessary) and repeat the above step–only this time hold the board vertically against the fence to make the second cut ❹. Make sure that the waste is on the outside of the blade, not on the inside against the fence where it could potentially cause a kickback or become a projectile.

3 Holding the board tight to the fence and table make first cut.

4 Reset blade and fence and make second cut with board held vertically.

JIGSAW

AN ORBITAL JIGSAW ALLOWS YOU TO CUT CURVES WITH EASE. IT WILL ALSO HANDLE STRAIGHT CUTS WHEN GUIDED BY A FENCE. AND IT'S ONE OF THE SAFEST TOOLS FOR MAKING QUICK CUTS.

While any good quality saw offers little vibration and smooth cuts, you'll appreciate a saber saw that features orbital blade motion The difference between standard and orbital motion will become apparent as soon as you make the first cut. Orbital action produces a much faster and cleaner cut than the standard straight up and down motion. Since blade selection makes a huge impact on performance be sure to use the manufacturer's recommended blade for the intended application.

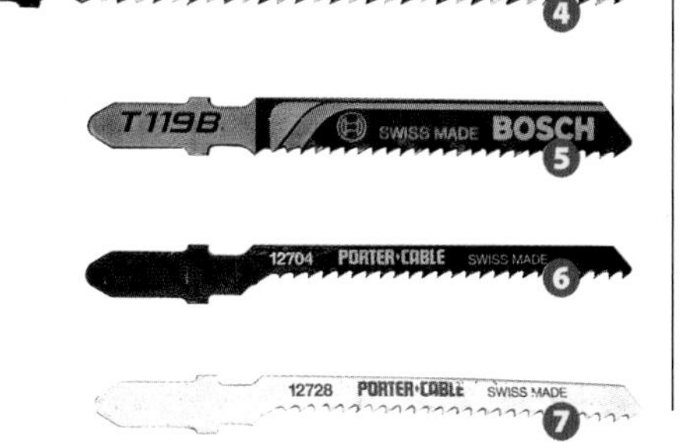

JIGSAW BLADES

1. Use for rough cuts
2. Use for general cutting purposes
3. Reverse tooth design reduces splitting
4. Bimetal for plastic
5. General purpose wood/ plywood
6. Narrow blade for curves
7. Cuts curves in plywood

USING AN INVERTED JIGSAW

1 On ninety-degree cuts keep blade perpendicular and base tight to board.

2 For precise beveled cuts change the base angle and keep tight to material.

3 When creating a back cut, tilt the saw holding one side of saw base tight to material.

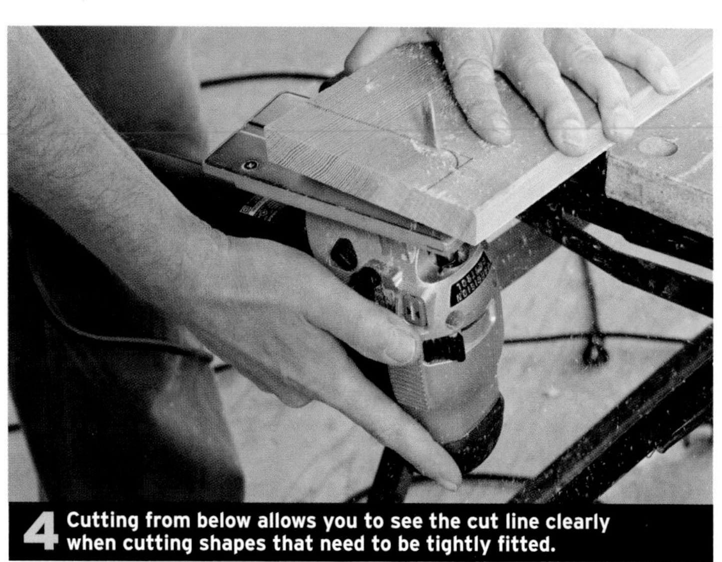

4 Cutting from below allows you to see the cut line clearly when cutting shapes that need to be tightly fitted.

While a saber saw is typically laid on top of the work to make a cut, I invert my saber saw to cut from below the majority of the time, finding this method to be more accurate and efficient. It may take a bit of experimenting before you are comfortable with this but it is well worth the effort. When working inverted, it's important to counteract gravity and keep the table of the saw tight to the bottom of the material ❶. If the cut needs to be at an exact angle, set the saw's table to that angle and hold the table tight to the bottom ❷. Alternately, if you are making a non-critical back cut simply tilt the saw and keep one edge of the table tight against the board ❸. The inverted method allows you to keep the blade in full sight next to the cut line with the added benefits that the majority of the sawdust comes out on the bottom side of the material and any splintering happens on the bottom as well ❹.

ROUTER

A FIXED-BASE ROUTER WITH 1 TO 1¾ HP OUTFITTED WITH THE RIGHT BITS CAN DO A WIDE VARIETY OF PROFILES.

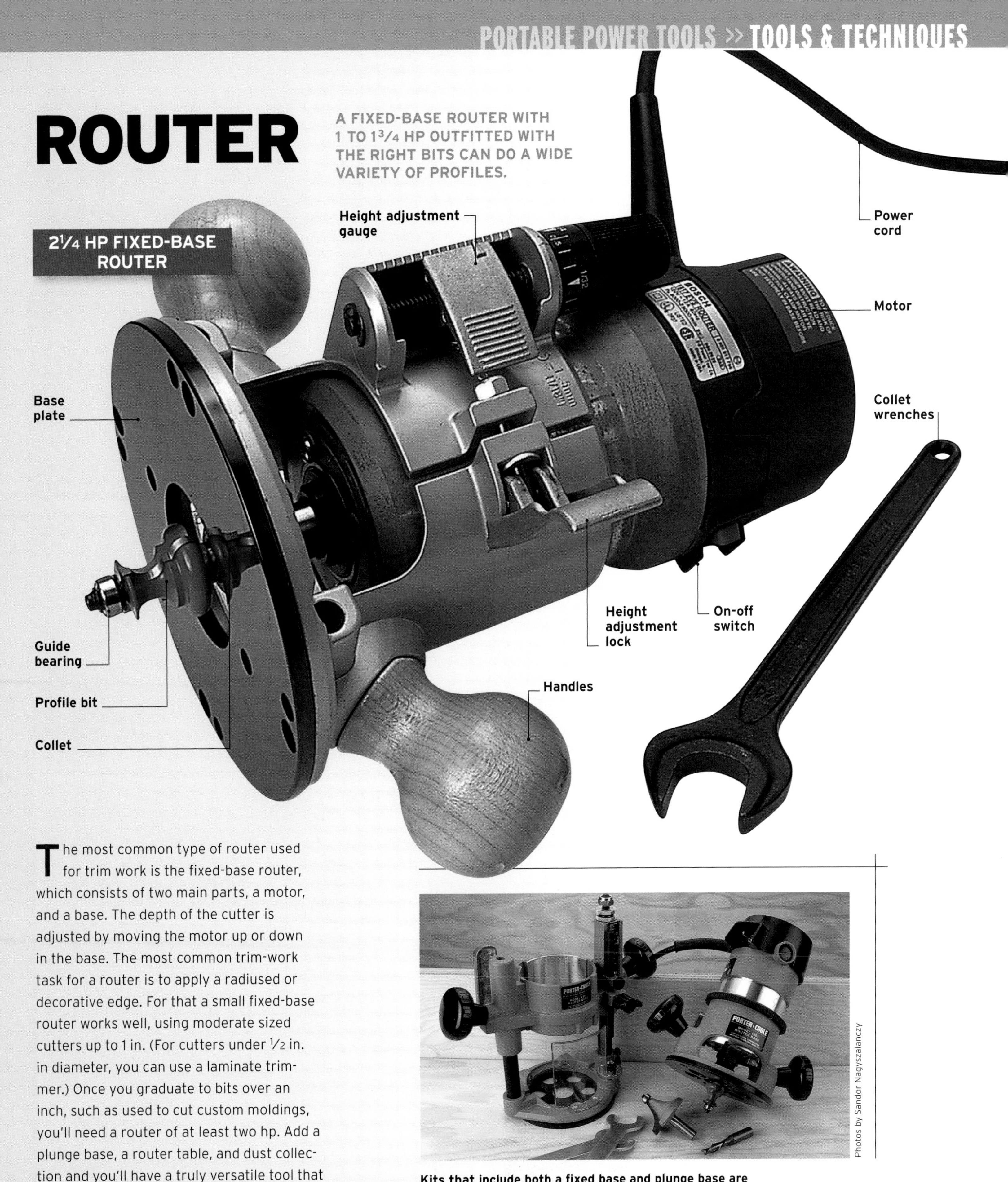

The most common type of router used for trim work is the fixed-base router, which consists of two main parts, a motor, and a base. The depth of the cutter is adjusted by moving the motor up or down in the base. The most common trim-work task for a router is to apply a radiused or decorative edge. For that a small fixed-base router works well, using moderate sized cutters up to 1 in. (For cutters under ½ in. in diameter, you can use a laminate trimmer.) Once you graduate to bits over an inch, such as used to cut custom moldings, you'll need a router of at least two hp. Add a plunge base, a router table, and dust collection and you'll have a truly versatile tool that can cut joints and do other decorative work.

Kits that include both a fixed base and plunge base are like getting two routers in one.

Photos by Sandor Nagyszalanczy

EDGE ROUTING TOPSIDE

Start the process by installing the bit into the router, inserting it into the collet and backing it off by 1/8 in. ❶ Securely tighten the collet nut ❷. Slip the motor back into the base and set the depth of the cut by adjusting the position of the router in the base and tighten the lock lever ❸. Body position is important when using a hand held router: Stand in such a way that you can maintain a nice steady feed rate while cutting. If you linger in one spot during the cut there is a good chance the cutter will leave a burn mark, which is difficult to remove. If you do have to stop, remove the router from the work immediately and then resume the cut when you get back in position. Always move the cutter into the material in the direction that the cutter is spinning. Hold the base flat to the material you are cutting ❹. When using a bit with a guide bearing, the bearing rubs directly against the stock, limiting the depth of the cut. If the bit does not have a bearing, use a guide to control the cut.

1 Insert the router bit and back it out approximately 1/8 in.

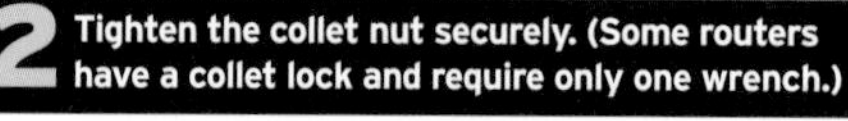

2 Tighten the collet nut securely. (Some routers have a collet lock and require only one wrench.)

3 Once bit is set to correct depth tighten the lock lever (or locking nut) to secure.

4 Hold router base flat on material using the guide bearing to guide cut.

Photos by Lonnie Bird

Router bits come in shapes and sizes to cut just about any profile you can imagine. By making multiple passes with a variety of profiled bits you may also able to create a custom profile or one that is too big to produce using a single bit.

FEED DIRECTION FOR HANDHELD ROUTER

ALONG ONE EDGE OF BOARD

Move the router so the wood feeds into the rotation of the bit.

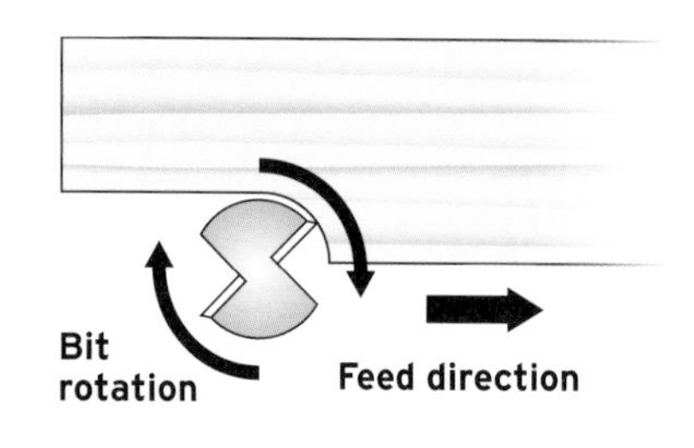

AROUND EDGES OF BOARD

Cut around the edges of a board. Start on an end-grain surface and follow with a long-grain pass to clean up any tearout.

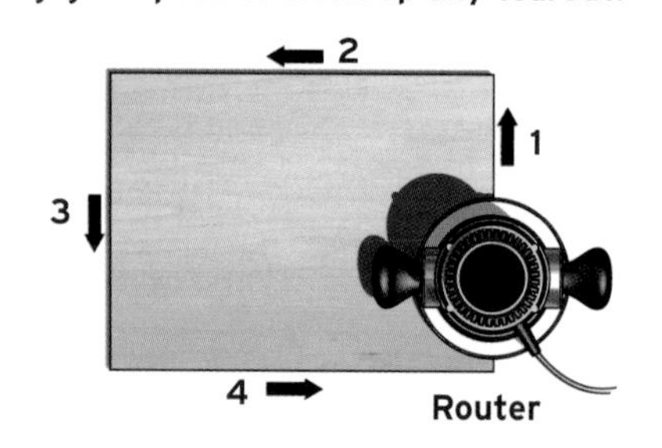

EDGE ROUTING ON A ROUTER TABLE

1 Set the height of the bit using a square as a guide. You can also set the stock beside the bit and estimate it by eye.

2 Using a straight edge to adjust the fence flush to the bearing.

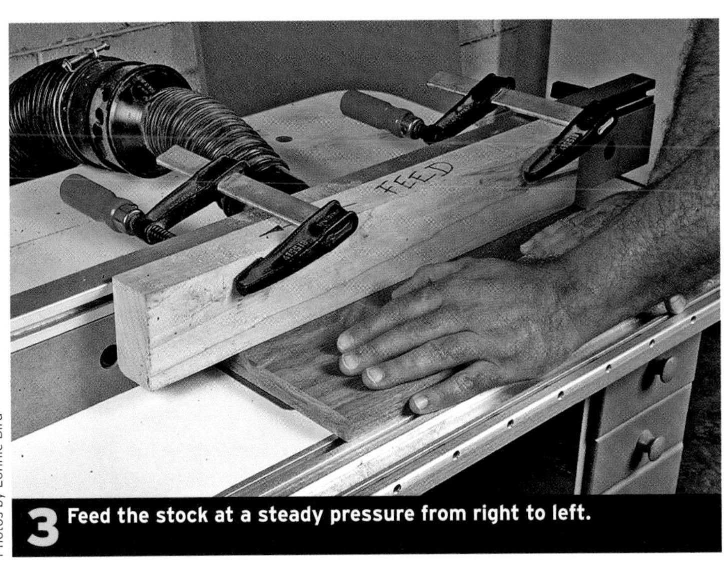

3 Feed the stock at a steady pressure from right to left.

Photos by Lonnie Bird

After securing the bit in the collet, set the router depth adjustment so the bit sits at the correct height above the table ❶. Using a straight edge, align the near side of the bearing with the front of the fence ❷. If the cutter has a guide bearing, a fence is not strictly necessary though the fence does add a margin of safety as well as a place to connect a vacuum hose for dust collection. Use the factory supplied guard or mount a piece of stock over the work to keep your hand from the cutter. When possible use feather boards to keep the material tight to the cutter.

On a router table you feed the stock into the router from the right to the left ❸. This pulls the material into the fence or the bearing to control the cut. Always feed the material into the spinning cutter at a slow speed using steady pressure. To adjust the depth of the cut and the amount of material being removed either move the bit up or down in the table or move the fence in or out from the bit.

FEED DIRECTION FOR ROUTER TABLE

On a router table, feed the workpiece into the bit from right to left, into the rotation of the bit. This will pull the work against the fence.

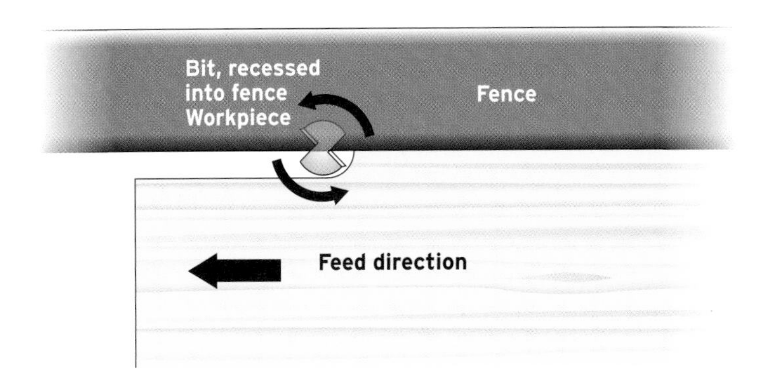

TRADE SECRET When making cuts using a large router bit do not attempt to cut the entire profile in one pass. Make several passes, gradually increasing the cut each time. This is easier on the router and the bit, produces a smoother cut, and prevents tearout.

CORDLESS DRILL DRIVER

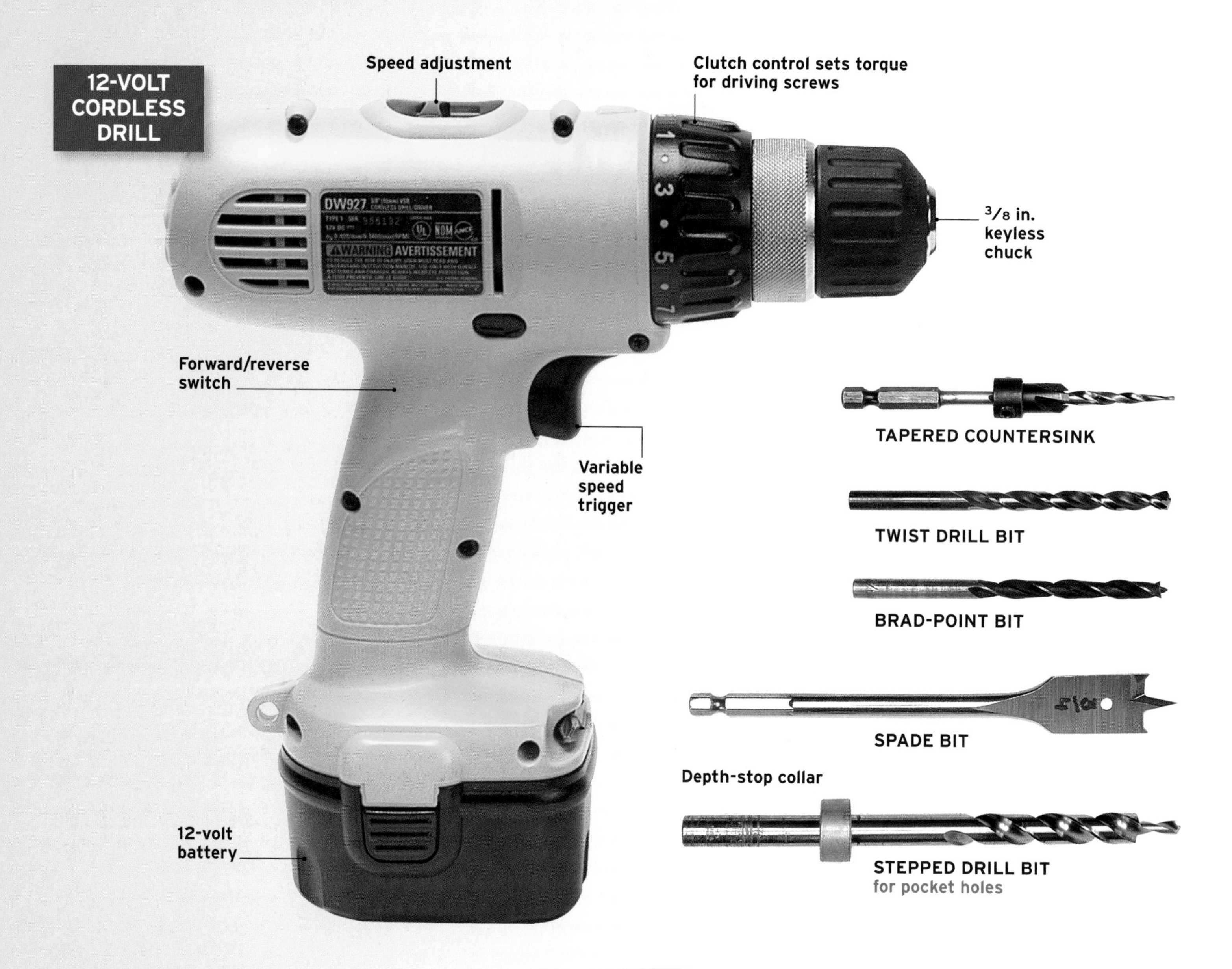

A good 3/8 in. drill is indispensible for certain kinds of trim installation. Look for a well-built drill but keep in mind that the battery is the key component. Impact drivers are a close cousin and are designed to drive threaded fasteners. Though annoyingly noisy, they are fast and spare your wrist from all the torque associated with screwing large fasteners while using a drill. When choosing a battery-powered drill for trim work, a lower voltage model is often a better choice as you don't really need the additional power of an 18-volt drill and you can certainly do without the weight. Rather than voltage, look for a drill with a battery that has a high amp hour rating. This is an indicator of how long the battery will last before requiring a recharge.

There are three common types of drill bits used to install trim. Although designed for metal the twist drill works in wood but is limited in the sizes available. Spade bits are available in larger sizes, cut fast and are easily sharpened on the job site. The brad point bit (which features a point that centers the bit when starting the hole) is used when very clean holes are required.

Conceal screws in a trim installation, by creating countersunk holes with a specialized bit. These bits form a tapered pilot hole for the screw while at the same time boring a hole to accept a wood plug over the top of the screw. You can buy plugs or make your own with plug cutters and drill press.

Photos by Phillip Dutton

INSTALLING POCKET-HOLE SCREWS

Pocket screw joinery uses a special stepped drill bit to simultaneously drill a pilot hole at a very shallow angle for a screw and a clearance hole for the screw head so that you can draw two workpieces securely together.

The bit utilizes an adjustable stop to limit the depth of the hole to account for the thickness of the stock.

➔ **See the drill bits on the facing page.**

The pieces being drilled are clamped into a jig that guides the drill at the correct angle and depth ❶. To assemble the joints after drilling, align the pieces on a clamping jig or simply clamped them to a flat surface. Now drive the specialized self-tapping screws into the holes and draw the pieces together. There is no need to clamp the stock together—the screws do the work ❷. Pocket screw joinery is typically used on the backside of components where the screw hole will be hidden. If, however, the screw hole will be visible, special plugs are available that fit into the angled hole ❸.

Photo by Phillip Dutton

The self-tapping screw enters the wood at an angle grabbing long-grain rather than weak cross grain. The screw shoulder seats at the bottom of the hole.

1 Clamp the board into the drilling jig and drill the pilot holes.

2 Use a clamping fixture to hold the boards and insert the screws.

3 When necessary the holes can be plugged with available angled wood plugs.

BISCUIT JOINER

1 Choose the biscuit size based on the width and thickness of the stock. Remember that the slot is always longer than the biscuit.

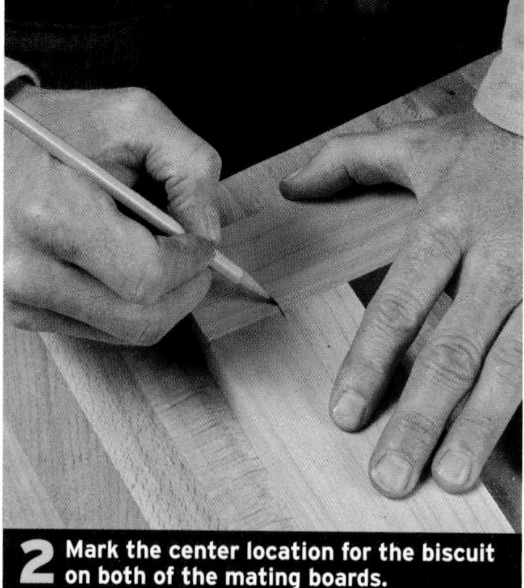

2 Mark the center location for the biscuit on both of the mating boards.

3 Align indicator on mark. Hold the biscuit joiner firmly against the board and push the body of the machine forward.

4 Repeat the process for the mating board.

5 Apply glue to both sides of the joint, working it into the slot. Center the biscuit and push it into the slot.

6 The joint before being clamped.

In biscuit joinery a pressed beech wafer acts as a loose tenon and is inserted into slots cut into adjoining boards. Biscuits are compressed beech and swell when they come in contact with water-based glues forming a tight joint. Standard biscuits are available in three different sizes ❶. In trim carpentry, biscuits are primarily used to provide face alignment when joining material together. Cut the two boards to be joined so they are square and then butt them together, marking the location of biscuits on both boards ❷. Then using the height adjustment on the biscuit joiner fence set the cutter to the center of the board. Align the top locator mark with the location marks on the boards. Hold the fence tight to the board, and press the switch. Make sure the cutter is running before pushing the body of the machine forward to cut the slot. Cut at a slow steady rate until the cutter reaches the preset depth ❸. Repeat the process for the mating board ❹. Clear any chips from the slot before gluing. Apply glue to both the board edges, working it into the slot. ❺ Insert the biscuit into the slot, keeping it centered ❻. Clamp the boards until dry.

You can also use this process for joining miters by cutting the slot into the miter.

Biscuits can also be used to reinforce miter joints.

Photos (left) by Gary Rogowski

PLANERS

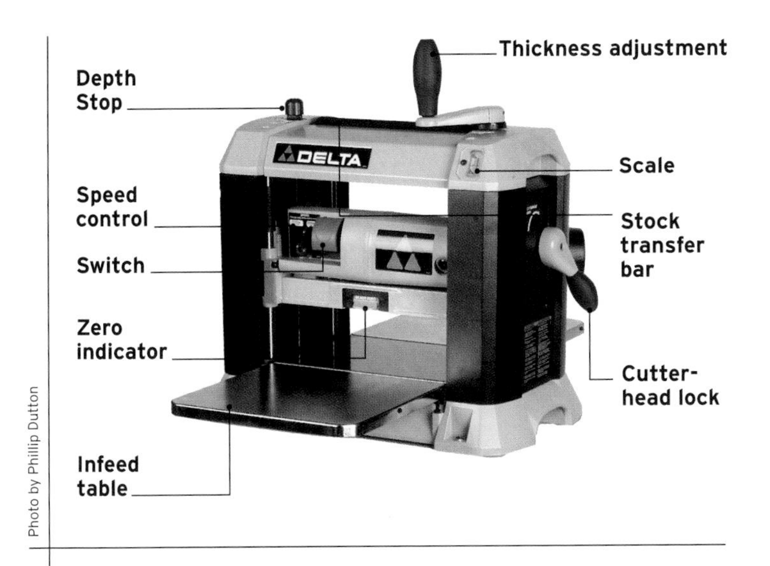

Portable thickness planers allow you to reduce the thickness of solid stock materials. These planers produce a very smooth finished cut and feature easy-to-change blades. When using portable planers don't remove more than 1⁄16 in. at each pass. To plane long stock, use support, such a roller stand set at the proper height, in addition to the machine's outfeed table **A**. Or, you can ask a helper to catch the stock as it comes out of the machine.

Planing thin stock

When planing stock less than 3/8-in. thick use a shop-built height fixture to create a platform that is raised off the bed of the machine **B**. This provides room to clear the chips that otherwise tend to plug up the planer when the cutter head is too close to the bed. Make the fixture by cutting a piece of plywood or melamine several inches narrower than your planers cutting width and long enough to go past the ends of the machine's infeed and the outfeed supports. Attach a cleat to the bottom of the platform at one end. Hook the cleat over the end of the infeed table to keep the fixture from being pulled through the planer with the material.

A power plane removes material quickly. Look for one with a dust collection port.

Power plane

Portable handheld power planes are used for cleaning up edges of material, straightening boards, and for planning stock in place. Look for a model that combines light weight with adequate size (most are at least 3 in. wide) and adjustable cutting depth. Note that some models have blades capable of being resharpened while others use throwaway blades. These tools make a lot of shavings fast so a dust port for a vacuum connection is definitely a worthwhile feature.

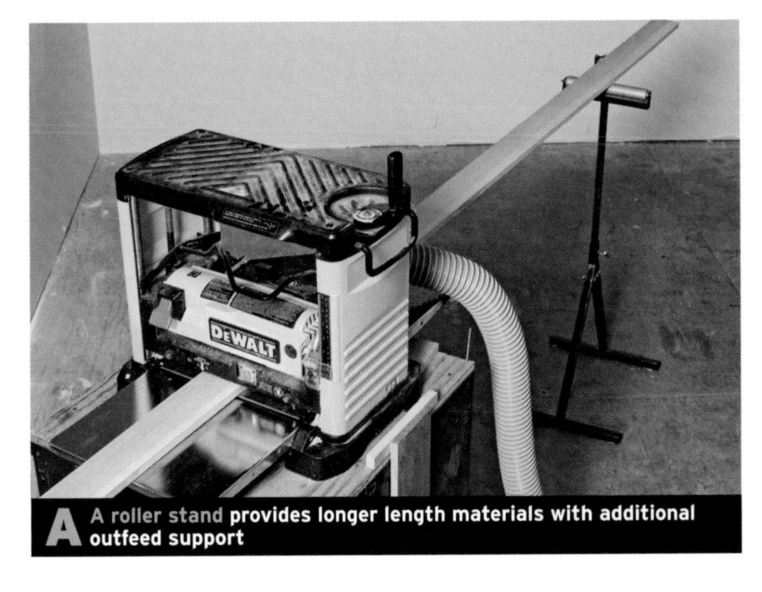

A A roller stand **provides longer length materials with additional outfeed support**

B A shopmade fixture **prevents the planer from clogging with chips when planing thin stock by creating clearance for the chips.**

NAIL GUNS

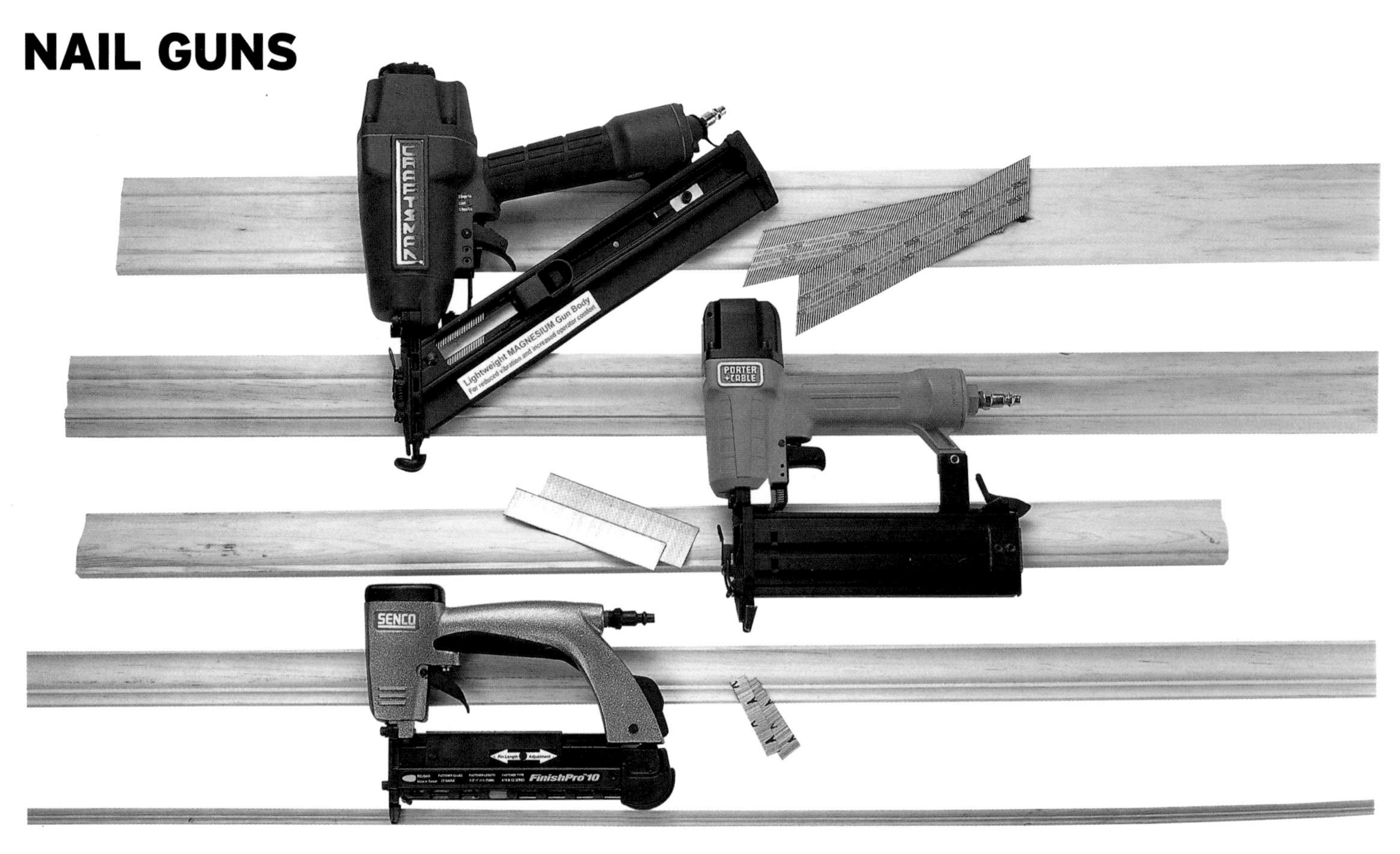

Choose a nailer that is appropriate for the job. A 15 ga. finish nailer (top) is for larger moldings, while the 18 ga. brad nailer (center) handles smaller and thinner trim. The 23-ga. (bottom) is for very fine moldings and hardly leaves a mark on the surface.

Compressors and nail guns used to be the tool of the professional. Today, they are so reasonably priced that most everyone can own one. Stay with a well-known brand and you will get a quality product. To power the guns, a small, quiet, and lightweight compressor unit (mine weighs only 26 lb.) is entirely adequate and will handle two trim carpenters day after day with no fuss. Nail guns have also become lighter and more compact over the years and are available in various styles that shoot a broader range of fasteners.

The most common nails come in 15, 18, and 23 gauges—the smaller the gauge the heavier the nail. The most universal size is the 18 gauge, which is considered a pin instead of a nail. They are adequate for installing most trim although they are somewhat small when used for installing crown and when hanging doors. Here it is best to use a 15-gauge nail. The headless 23-gauge nail is called a micro pinner and is used exclusively on small moldings.

WARNING Remember to treat a nail gun as you would a firearm. Never point it at anyone or disable the safety.

A small "pancake" compressor is adequate for most trim jobs.

Photo by Sandor Nagyszalanczy

SANDERS

Nobody likes to spend a lot of time sanding, and with the sophisticated technology of today's electric sanders you really don't need to. The old standby was an oscillating sander that produced a great finish, but did so very slowly. Then the random orbit sander came along and changed the sanding world. These new machines are capable of quickly removing a lot of material while producing a silky smooth finish free of sanding marks. Belt sanders are the serious material removers and range in size (measured by belt width) from 1 in. to 4 in. For trim work the smallest varieties are invaluable for making radiuses or trimming to a scribed line. To control the huge amount of dust power sanders produce make sure the model you buy can be hooked up to a vacuum.

Dust collection

Any kind of dust is bad for our health but with the popularity of MDF and other man-made wood substitutes dust control is a necessity. Always use a dust mask—or purchase a visored helmet fitted with a powered filtration system if you find the dust to be excessive or bothersome. You'll find that the dust bags supplied with most tools are not adequate except for the lightest, occasional work. Instead, hook the tool's dust port up to your shop vacuum.

Some tools (like planers) generate amounts of dust and shavings too great for a shop vacuum to handle. Dust collection systems are available that are very portable and easily set up.

A vacuum works to collect the dust generated by most portable power tools.

Photo by Sandor Nagyszalanczy

Power sanders make quick work of sanding tasks: (clockwise from top left) A 4-in. disk grinder, a 1/4 sheet palm sander, a belt sander, and a random orbit palm sander.

Photo by Jim Dugan

HANDSAWS

A Panel saw **works well for making rough cuts and breaking down long pieces of lumber.**

B Western-style backsaw **is used for finish cuts and joinery.**

While the process of cutting a board with a handsaw is admittedly slow going and usually produces a rougher cut than that of a power saw it is sometimes a better or more appropriate choice. A handsaw can rough cut lumber to length or finesse small cuts. They are also much less noisy than power saws and produce less dust. For rough cutting to length or for cutting framing material you can use a traditional western hand crosscut saw **A**. There are also saws this size that have Japanese style teeth that are faster cutting but cannot be resharpened. When the blades dull, they must be replaced. For making finish cuts, a fine-toothed backsaw produces the best results. While the traditional western style backsaws work fine, it produces a wider kerf **B**. A Japanese style pull-saw cuts faster and more smoothly **C**. For coping inside corners use a coping saw. Be aware that the blade cuts on the pull stroke toward the handle **D**. A general use saw for cutting sheetrock or rough stuff is a saw handle that uses replacement reciprocating saw blades **E**. The blades are easily interchangeable and the saw will fit in your apron.

C Japanese-style backsaw **cuts on pull stroke and has very thin kerf.**

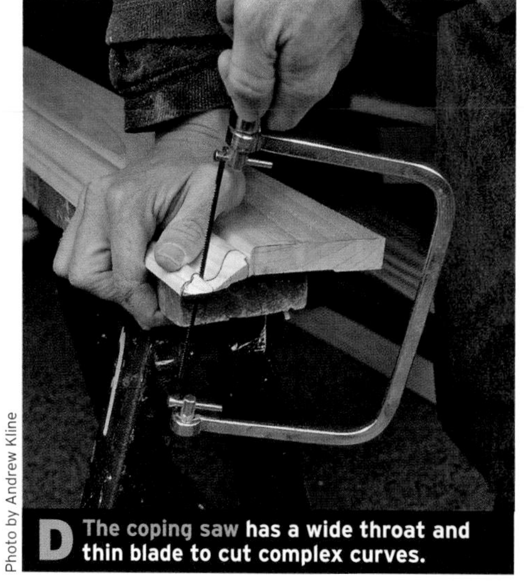

Photo by Andrew Kline

D The coping saw **has a wide throat and thin blade to cut complex curves.**

E General-purpose saw **handle uses replaceable reciprocating saw blades.**

CUTTING MITERS ON A HAND MITER BOX

A hand-operated miter box may be slower than a power miter saw, but it is still a very effective method of cutting accurate miters. The major angles all have detents to set the saw accurately and some varieties have provisions for cutting compound miters. Modern versions of this tool use a tensioned replaceable blade much like a hacksaw. As is true with all handsaws the hand miter box will produce a superior cut if you let the saw cut at its own speed without applying too much downward pressure. To make a crosscut, lift the saw and slip the stock under the blade aligning the cut mark to the blade. Start by pulling back on the saw to create a starting groove on the material. Now push the saw forward and continue until the stock is cut. To cut a simple miter, hold the material against the fence and set the miter scale to the desired angle **A**. To cut a compound miter, hold the stock at an angle between the table and the fence and lock the saw guide at the desired miter setting **B**. A simple miter box can also be used for small moldings, but it's best reserved for quick jobs that require only a few cuts. Simple miter boxes have a tendency to become inaccurate with repeated use and should be replaced when they begin to wear.

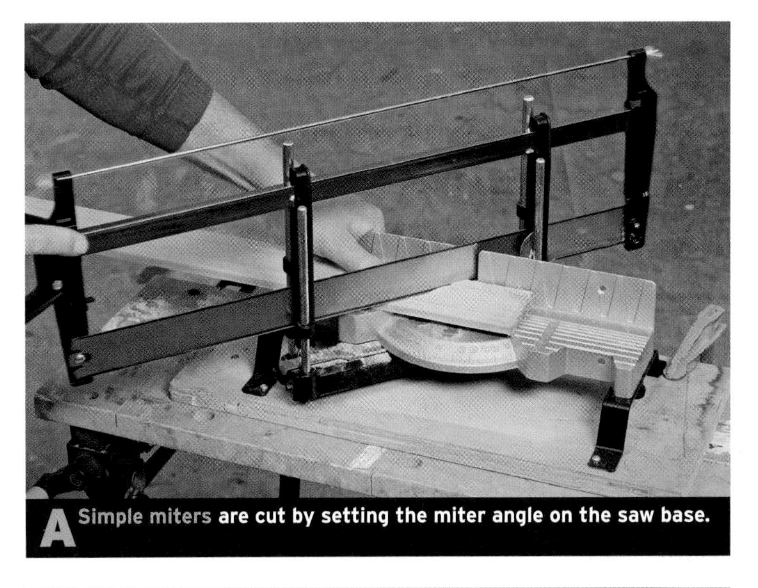

A Simple miters are cut by setting the miter angle on the saw base.

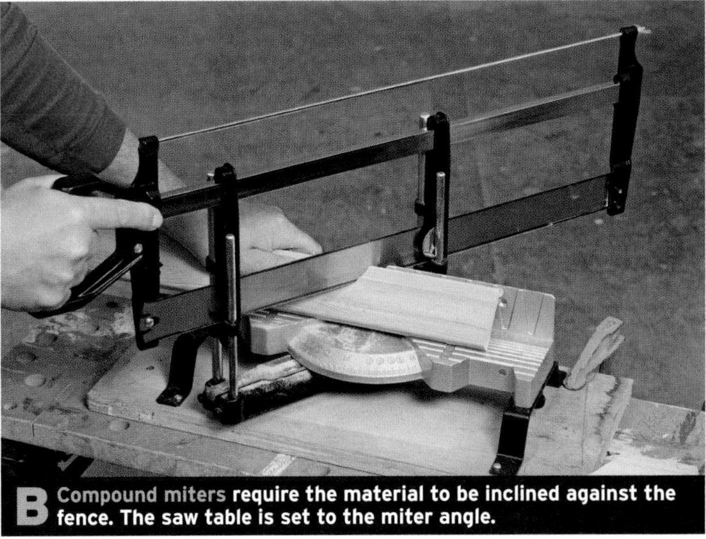

B Compound miters require the material to be inclined against the fence. The saw table is set to the miter angle.

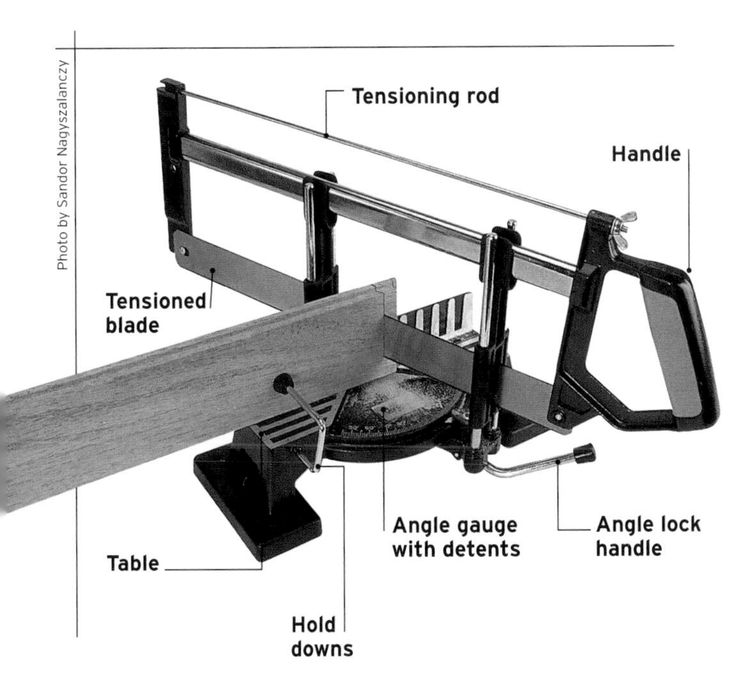

Photo by Sandor Nagyszalanczy

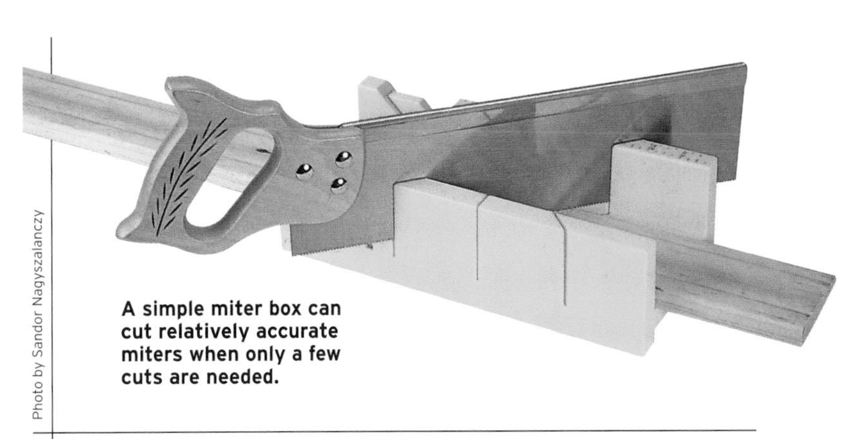

A simple miter box can cut relatively accurate miters when only a few cuts are needed.

Photo by Sandor Nagyszalanczy

COPING AN INSIDE CORNER

Coping inside corners of complex crown molding is sometimes best done with a coping saw. Begin by cutting a reverse bevel and miter on the end to be coped ❶. The reverse miter clearly shows the profile, but you can run a pencil over the edge to help define it visually. Begin cutting the profile, working from the center out so the saw has something to bite into as you start the cut ❷. As you remove the material, you are backcutting at 90 degrees ❸. It helps to saw from different directions, depending on the shape of the profile, to avoid breaking the edge ❹. Once you have sawn as close to the profile as possible, you can refine the cut, if necessary, with sandpaper or a rasp ❺. The finished joint has a continuous, smooth edge ❻.

Finding the profile is easy when you use a pencil to darken the edge. Especially on moldings that haven't been primed, it's sometimes difficult to see the exact edge of the profile. It all looks like wood. The visual reference of a dark line can help you the saw the edge more accurately.

1 Cut a reverse bevel to expose the profile.

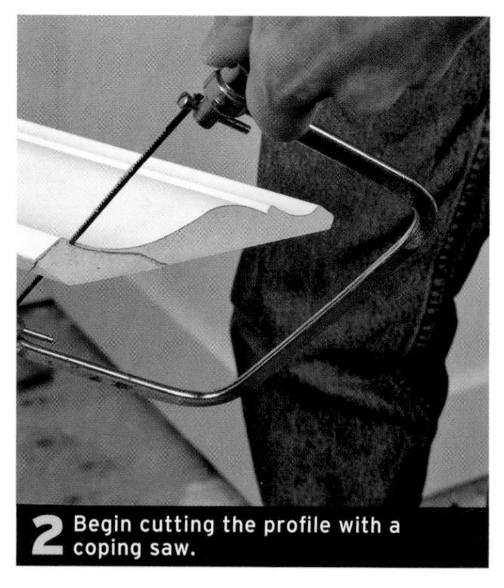

2 Begin cutting the profile with a coping saw.

3 As you cut, backcut by 90 degrees.

4 Work from different directions to avoid breaking the edge.

5 If necessary, smooth with a rasp.

6 The final profile.

Photos by Andrew Kline

CHISELS AND PLANES

A well-sharpened plane or chisel is often needed to fine-tune a cut or to make cuts where a power tool just can't reach. The key to good performance is to keep your edge tools sharp and protect the blade edge from damage.

Chisels

The chisels in a finish carpenter's toolbox live a hard life. Their handles are continually beaten and their edges can be dulled or damaged by a hidden nail and other hazards. In spite of the hard use, I have used this same set of chisels for 20 years. They were a mid-priced set with good steel and a plastic handle that is virtually indestructible. A set of chisels is usually a better purchase than buying chisels individually. Reserve your best chisels for cutting and paring tasks. Use a pry bar for trim removal, and keep some inexpensive "beaters" in your toolbox for use where you think you could encounter nails.

See "Pounding and Prying Tools," p. 46.

Planes

The most often used hand plane when installing trim is the block plane. Its small size allows one-handed use and offers a level of control not available with other larger planes. Unlike other planes, a block plane blade is mounted bevel up. The low angle versions excel at trimming the ends of mitered millwork. An adjustable throat plane allows you to cut very fine shavings.

As the planning tasks grow larger so do the planes. For general straightening and smoothing of boards a mid-size bench plane is useful. The plane shown here is an 04 smooth plane. For best results, secure the material the material to be planed on a workbench.

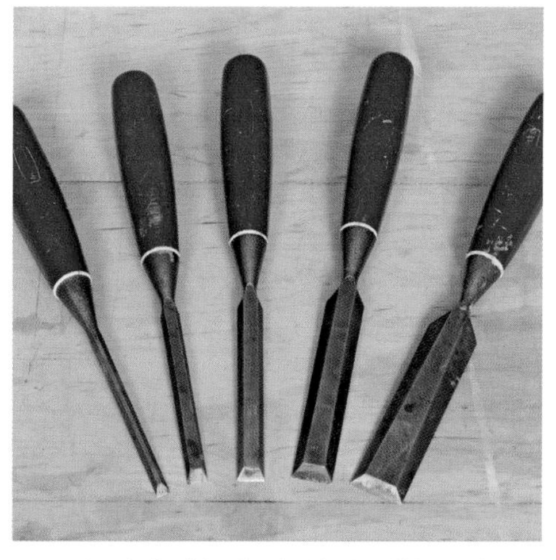

A good set of mid-priced chisels with quality steel for general work

Low angle block planes excel at small work like trimming miter joints

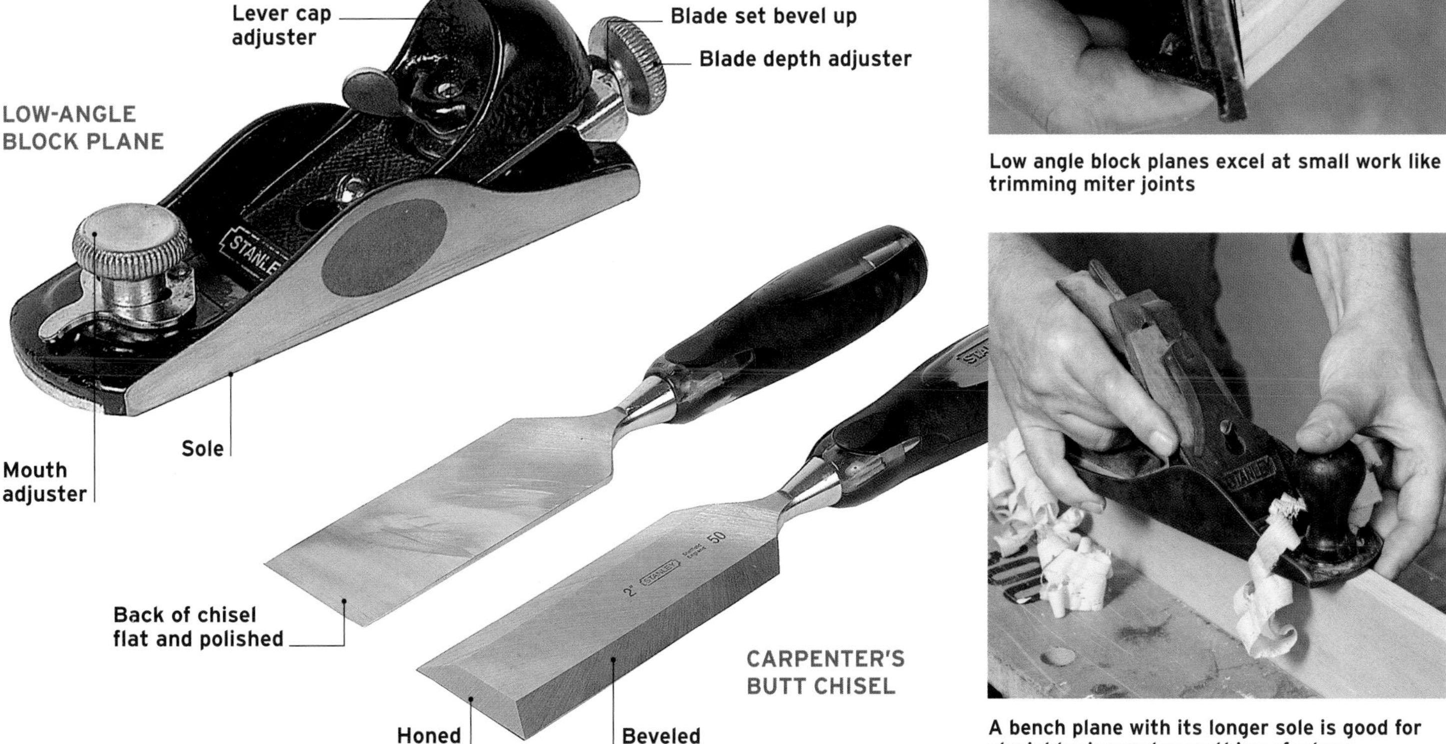

A bench plane with its longer sole is good for straightening and smoothing of edges.

POUNDING AND PRYING TOOLS

To someone who uses one a lot, hammers are as individual as a pair of shoes. And, like shoes, hammers are available in a plethora of different styles, weights, and materials. While the curved claw provides more leverage for pulling nails, the straight claw is more versatile. Fiberglass or metal handles are very strong and seldom break while pulling a nail.

A wood handle, however, offers much better shock absorbing ability and weighs less—keeping more weight focused at the head where it belongs. While there are many hammer weights to choose from, you will seldom need anything heavier than 16 oz. for installing trim. You only need enough force to drive the nail, so it make sense that a lighter hammer is used to drive finishing nails. I broke my right wrist some years back and when I resumed work I was forced at first to use a light hammer. I chose a 10 oz. straight clawed Vaughn and am still using it

Prying tools have improved a lot since I started carpentry. The combination pry bars that have a blade on one end and a nail puller on the other are the most useful for trim installation. The 6-in. version fits nicely in a nail bag and has a thin blade that is better for getting under and behind millwork. The best versions of these are made in Japan. A flat bar is used where a bit more force is needed and is meant for heavier work. Nail pullers known as cats' paws come in different sizes and are used whenever any serious nail pulling has to be done. They have a nail pulling claw at each end and are driven under the nail with a hammer when used. A good set of nail sets in different sizes is handy to have to recess nails below the surface or on the occasion when a nail gun doesn't drive the nail deeply enough.

A lightweight straight claw hammer is an essential in any trim carpenter's tool kit.

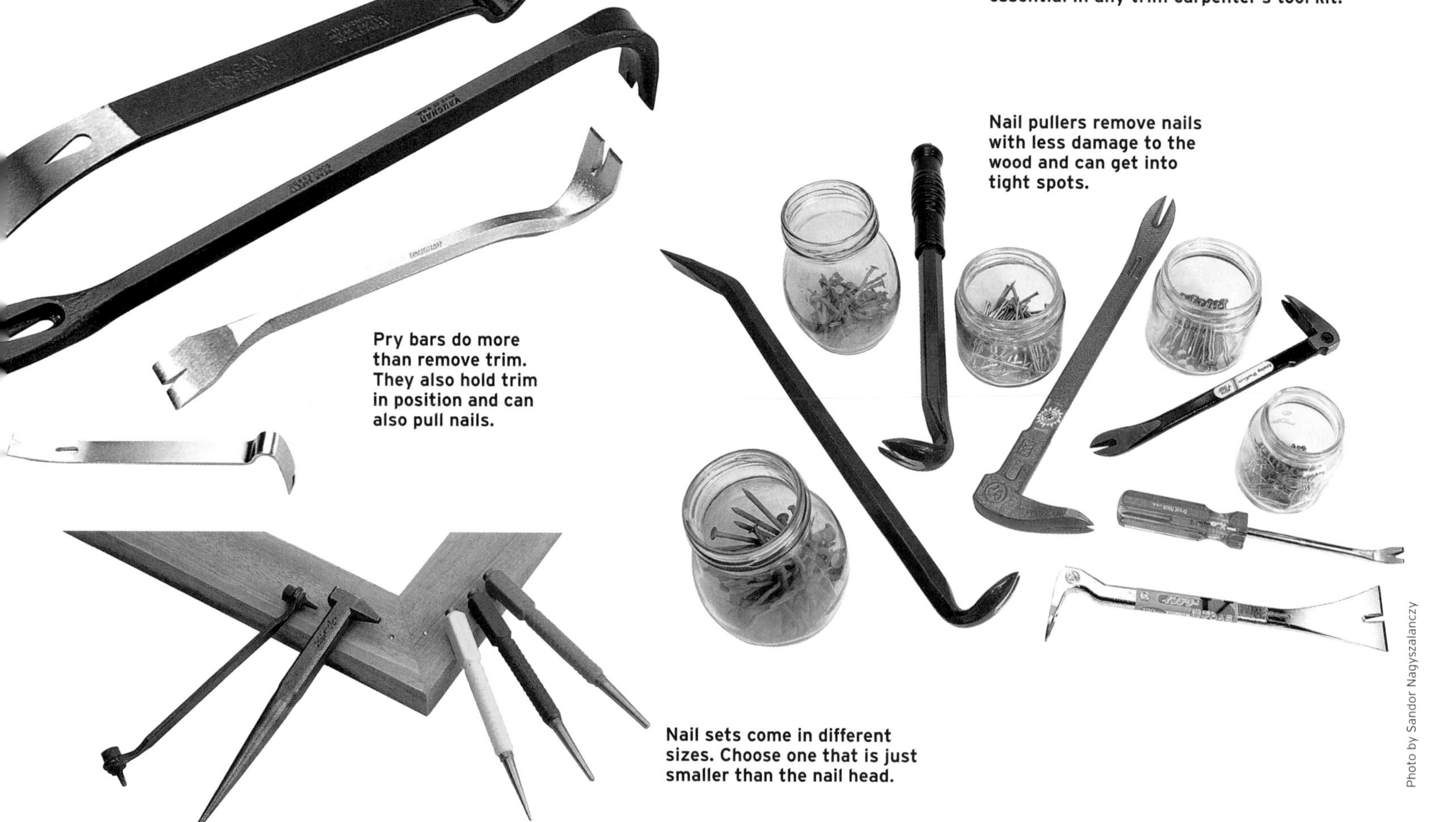

Pry bars do more than remove trim. They also hold trim in position and can also pull nails.

Nail pullers remove nails with less damage to the wood and can get into tight spots.

Nail sets come in different sizes. Choose one that is just smaller than the nail head.

Photo by Sandor Nagyszalanczy

MARKING AND MEASURING TOOLS

Accurate measuring and layout tools play a vital part in getting quality results for your efforts. It doesn't pay to skimp here. A square that isn't square is worthless.

Tapes and rulers

Retractable tape measures are the most common tool for determining length, especially for cutting stock to rough lengths. Buy one that's accurate and easy to read and long enough to go the maximum length of the room you'll be working in. Stick to one ruler or tape rather than switching off from one to the other. One ruler's inch may not be another's. For the same reason, I prefer to mark the length directly onto the stock or use pinch sticks to avoid measuring and the potential for error while reading and transferring the measurement. As a general rule, when doing complicated layouts take all the measurements from the same points whenever possible and do not take one measurement off the last, which can compound any previous error.

Squares

The most commonly used squares while installing trim are the framing square, Speed Square, combination square and sliding bevel gauge. A framing square and a 4-ft. sheet rock square work well for laying out plywood paneling. The obvious use of a square is to make sure that a layout or workpiece is square. Specialized squares can do more than measure 90 degrees. A sliding square allows you to transfer measurements to several work pieces without re-measuring. A Speed Square can serve as a fence for crosscutting with a circular saw. A sliding bevel allows you to transfer angles without the need to measure the degrees.

Marking tools

While it might seem obvious, a well-sharpened pencil is an important marking tool in trim work. A dull pencil leaves a thick mark and room for error. It also helps to mark the waste side of the line so you remember on which side of the mark to cut. Combine a pencil with a compass and you have a scriber that allows you to transfer irregularities of the wall or floor to the workpiece. A compass also allows you to draw circles and arcs, as well as divide a line into equal segments. Chalk lines are invaluable when you need to run trim, such as a chair rail around the perimeter of a room.

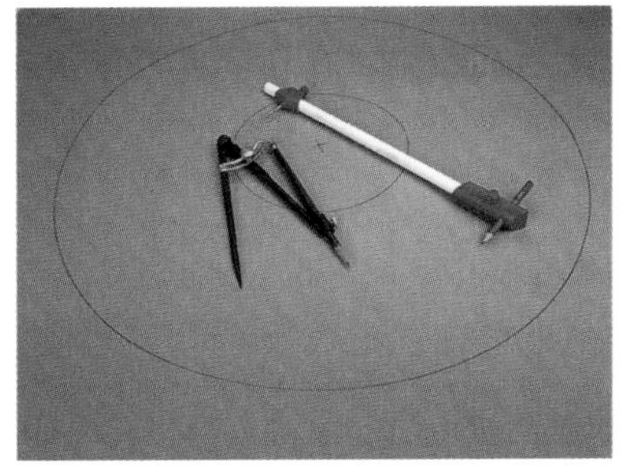

A compass can scribe irregular floors or walls for a tight fit.

Snapping a chalk line gives you a visible line for layout. On the left is a standard chalk line, on the right, a Japanese ink line which makes thin, highly accurate marks.

Photos by Sandor Nagyszalanczy

LEVELS

Perimeter moldings such as chair rails and perimeter bands have to be level or you'll have an unhappy situation where they meet. And there are many other tasks where a level is essential, such as installing doors. You'll need levels in several lengths. A journeyman's level of 36 in. to 48 in. is good for installing double doors, but a smaller level of 18 in. is handy and can fit into tighter spots. Buy the best levels you can afford—you do get what you pay for. Laser levels are becoming more affordable and can be a great benefit on jobs that require transferring heights around the perimeter of a room.

TRADE SECRET If the need for a long level only comes up occasionally a long straight board of the needed length can be taped to a shorter level. To check the accuracy of a level, place it on a surface and check the position of the bubble. Now flip it over and place it in the same spot. If the level is reading correctly, the bubble will be in the same position. If not the level is out of calibration.

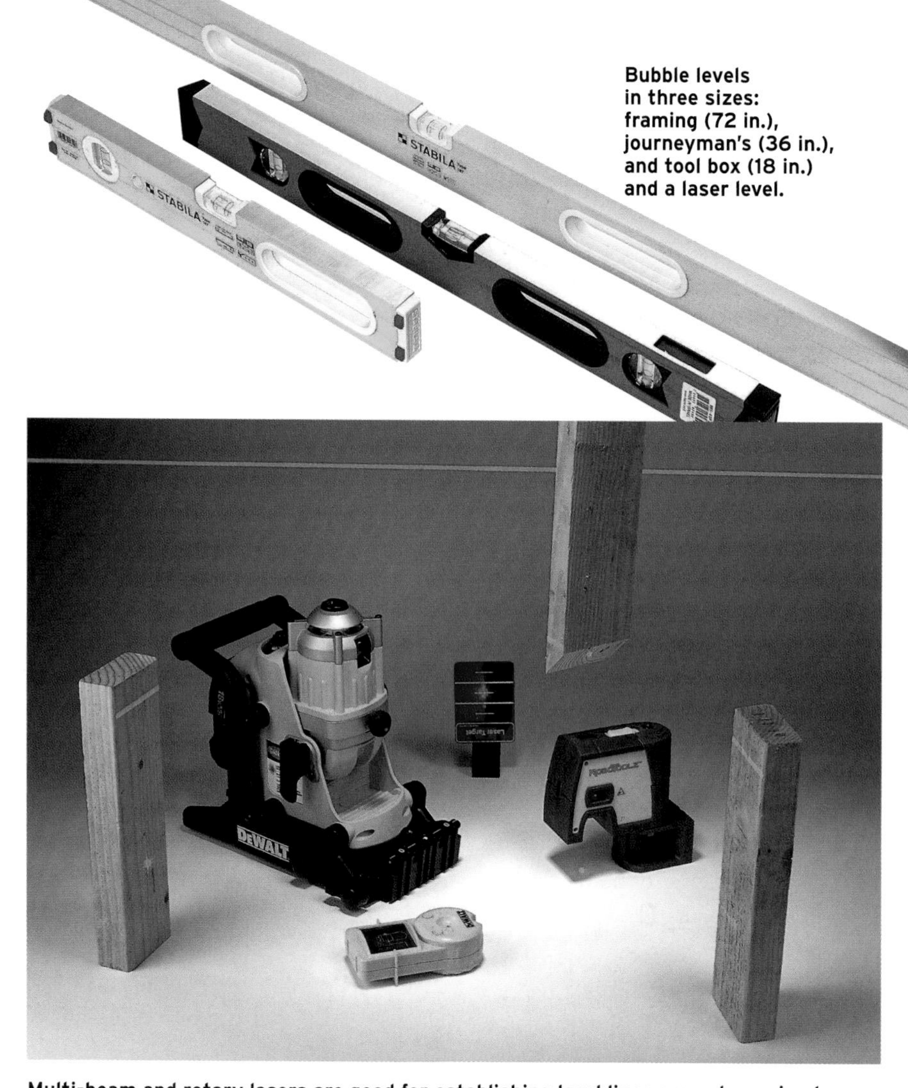

Bubble levels in three sizes: framing (72 in.), journeyman's (36 in.), and tool box (18 in.) and a laser level.

Multi-beam and rotary lasers are good for establishing level lines around a perimeter.

Photo by Craig Wester

A quality level is a necessity for installing doors and interior trim.

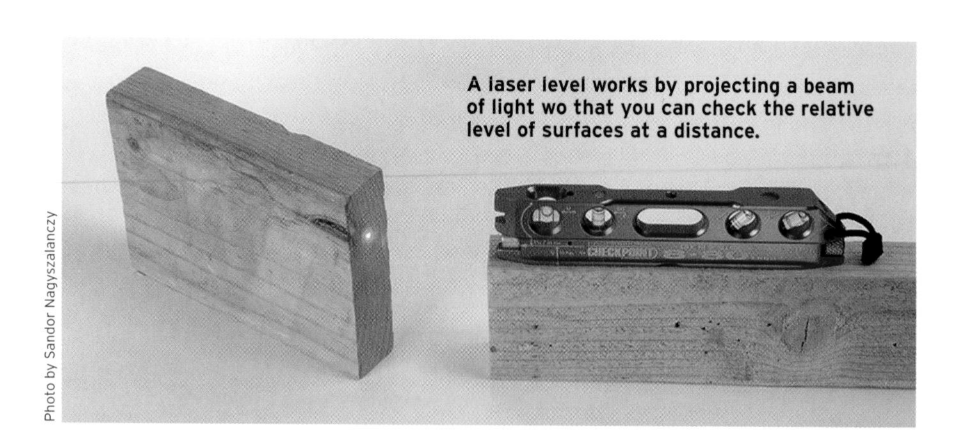

Photo by Sandor Nagyszalanczy

A laser level works by projecting a beam of light wo that you can check the relative level of surfaces at a distance.

DETERMINING ANGLES

There are a wide variety of tools and techniques to determine angles. The simplest and most common tool to determine and transfer angles is the sliding bevel (also called an adjustable bevel gauge). To use it to transfer an angle to a miter saw, simply hold the gauge to the fence and pivot the saw until the blade matches the angle. The actual reading in degrees can then be read off the saws miter scale **A**. To read the degree of angle directly from the work use a tool with the degree scale already on it **B**. Bisecting angles can be determined by striking arcs, using a proportional divider **C** or by using a combination tool like the Starret pro site angle finder, which gives readings for both single and miter cuts **D**.

Photo by Sandor Nagyszalanczy

Ways to measure and transfer angles: (from left to right) a sliding bevel, a miter square, and an electronic bevel gauge.

A Sliding bevel **transfers an angle to the chopsaw.**

B Protractor type angle gauge **determines an angle in degrees.**

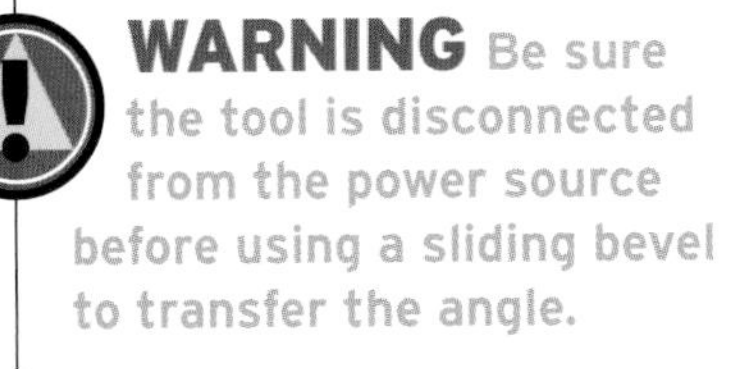

WARNING Be sure the tool is disconnected from the power source before using a sliding bevel to transfer the angle.

C A proportional divider **can determine bisecting angles.**

D This angle gauge **will give angles and their bisecting angle in degrees.**

TAKING INSIDE MEASUREMENTS

A A folding rule with a retractable brass extension allows you to measure the distance directly.

B Pinch sticks are butted to both ends and marked at overlap point.

C A slightly longer piece of wood than needed is held in position and marked to length.

When using a tape measure to get inside measurements it can be very difficult to accurately read the bent over tape in the corner. One way to eliminate this problem, butt the case to one end and add its length to the number indicated on the tape. For distances of less than 3 ft., a folding rule with a retractable brass extension works very well **A**. For very accurate work use pinch sticks **B** or, better yet, mark the piece in place **C**. Pinch sticks are butted to opposite sides of the opening and are either marked with a pencil where they overlap or secured in place using a clamp designed for the purpose. An option for longer distances beyond a metal tape's capacity to support itself is to mark a spot approximately midway between the walls and measure from each of the walls to the marked point adding the separate measurements to get the overall length.

REMOVING OLD TRIM

In trim jobs in older homes, you may have to remove existing trim before you can start. In some cases you may wish to save the trim for reuse. In that case, use a drift (a nail set with a long straight tip) to pound the nails all the way through the board ❶. Don't use a nail set as its tapered shaft tends to split the wood.

If the edges of the trim have been caulked to the wall, use a utility knife to cut through the caulk at the edge of the trim ❷. This will prevent the paper on the Sheetrock from tearing off when you remove the trim.

To avoid splitting the trim while using the pry bar to pull it away from the wall first, start prying by taking advantage of any existing gaps between the wall and the trim. Insert the pry bar into the gap and lever the trim away from the wall, working your way up and down the board until the board is free. Place a thin piece of wood behind the pry bar immediately next to the casing to give the pry bar more leverage and to protect the wall surface from being crushed by the pry bar ❸. In some cases, especially in older homes, there are no gaps between the wall and the trim with edges sealed under multiple coats of paint. Driving a pry bar—even a thin-bladed one—between the wall and trim will likely damage the trim significantly. Instead, drive a 1½ in. putty knife between the wall and the trim ❹. Keep working the tool up and down the casing until you create a large enough gap to insert the thin-bladed pry bar. Once the board is off the wall, remove any remaining nails with a set of pliers by pulling them through the backside. This minimizes damage to the face of the trim ❺. Driving the nail through from the back will usually result in the splintering of the face around the nail hole.

1 Use a straight shank punch to drive nail through trim to avoid splitting.

2 Use a utility knife to cut through caulking.

3 A piece of wood placed behind pry bar increases leverage and prevents damage.

4 Use putty knife to get behind trim too tight for pry bar paint.

5 Pull nail through from back of trim to prevent splintering of face.

MAKING BACK CUTS

A Use a tablesaw **to back cut a straight board.**

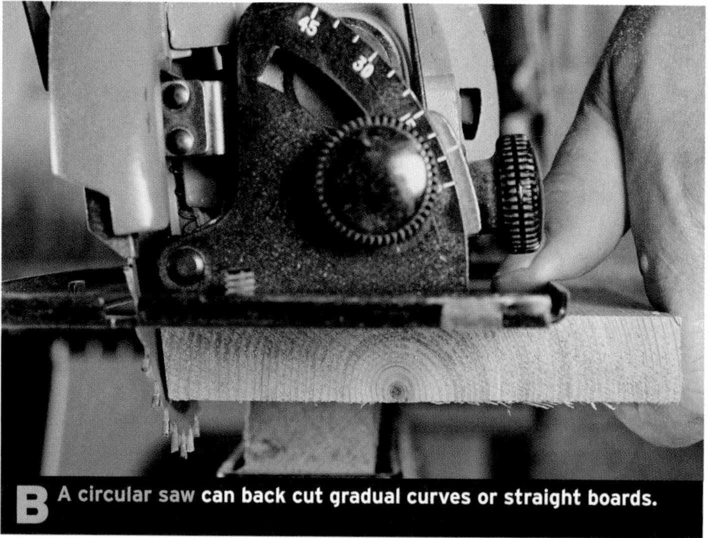

B A circular saw **can back cut gradual curves or straight boards.**

C A jigsaw **is the tool of choice for irregular cuts.**

D A belt sander **works for irregular cuts and cleaning up rough edges.**

When fitting the edge of a board to a straight or irregular surface, back cutting (which is simply undercutting) the edge often produces a tighter, better-looking joint. Undercutting ensures that the visible edge of the board fits tightly without any interference from the rest of the board edge. This is a cosmetic technique and should not be used in applications where strength is required or the joint will be subjected to stress. On a straight board with parallel sides the easiest way to create the backcut is to set up a table saw with the blade cranked several degrees away from perpendicular **A**. To back cut to an irregular surface, however, you can choose from one of the tools and methods illustrated below.

If the edge requiring the back cut runs at an angle along the length of the board, or if it runs at a very gradual curve, a circular saw is a good choice for making the cut. Set the blade to undercut at about a 5-degree angle and run it slowly and carefully along the cut line. Note that the smaller the diameter of the blade the tighter the curve that can be cut **B**.

If the cut line is highly irregular, the jigsaw is the best tool to make the cut. If your saw's table is adjustable, set the table to a 5-degree angle, place the body of the saw on top of the work with the blade protruding out below and follow the scribed line. An alternative (though a bit trickier) method is to make the backcut by running the saw's body below the face of the workpiece so that the blade protrudes upward to the "good" side **C**. The advantages of this method are threefold: Good visibility (as almost all of the sawdust falls away from the cut); improved control (because you can see the blade with no saw in the way); and a cleaner cut (as all of the splintering happens on the bottom, hidden side of the board due to the blade cutting on the down stroke).

E A block plane works well for back cuts and trimming to the line.

To make the cut, set the table to about a 5-degree angle (or simply tip the saw up on one side of its base) and, keeping the table tight to the underside of the board, run the saw along the length of the cut as you watch the progress of the blade (rather than the tool itself).

Though this is not the standard use of this tool, you can use a belt sander to work a back cut into a scribed edge. Any belt sander will work, but generally the smaller the sander the easier it is to sand to the line **D**. I prefer a belt of less than 1½ in. for detail work. For more gradual curves, I use a sander with a 3 in. wide belt.

Finally, a hand plane is a capable (and quiet!) tool for creating a back cut. The block plane's short length allows it to work well on inside curves and is easily held in one hand **E**. Longer planes are more appropriate for longer, straighter lengths of cuts.

WARNING

In some of these photos the saw blade is extended for clarity. Only expose the amount of blade you need to execute the cut.

WINDOWS

THE PRACTICAL JOB OF WINDOW TRIM is to cover up the gap between the wall frame opening and the window frame. Beyond the practical, window casing expresses architectural style, whether straightforward contemporary or elaborate Victorian. Window trim has high visibility—you see it every time you look out a window. So it's important that you select the appropriate style for your home and get the installation right with even margins all around and tight joints at all the corners.

The techniques presented in this section deal with all the components of window trim so you'll be able to install any style window trim successfully.

Because it's such a focal point in any room, well-executed window trim shows off good workmanship and fine materials. When choosing trim materials, consider whether the trim will be painted or have a clear-coat finish. Natural finishes look best when the trim components are matched for color and grain compatibility.

JAMBS

WINDOW STOOLS & APRONS

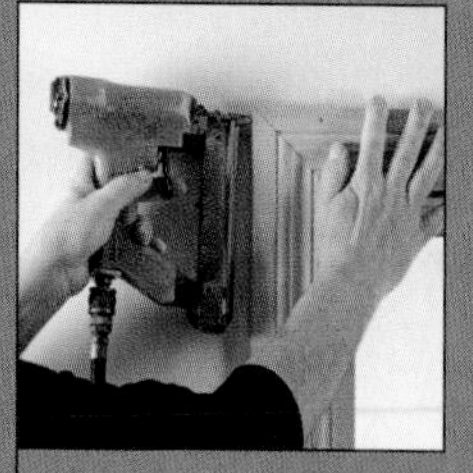

WINDOW CASING

WINDOWS IN A SERIES

WINDOWS WITH STYLE

WINDOW TRIM

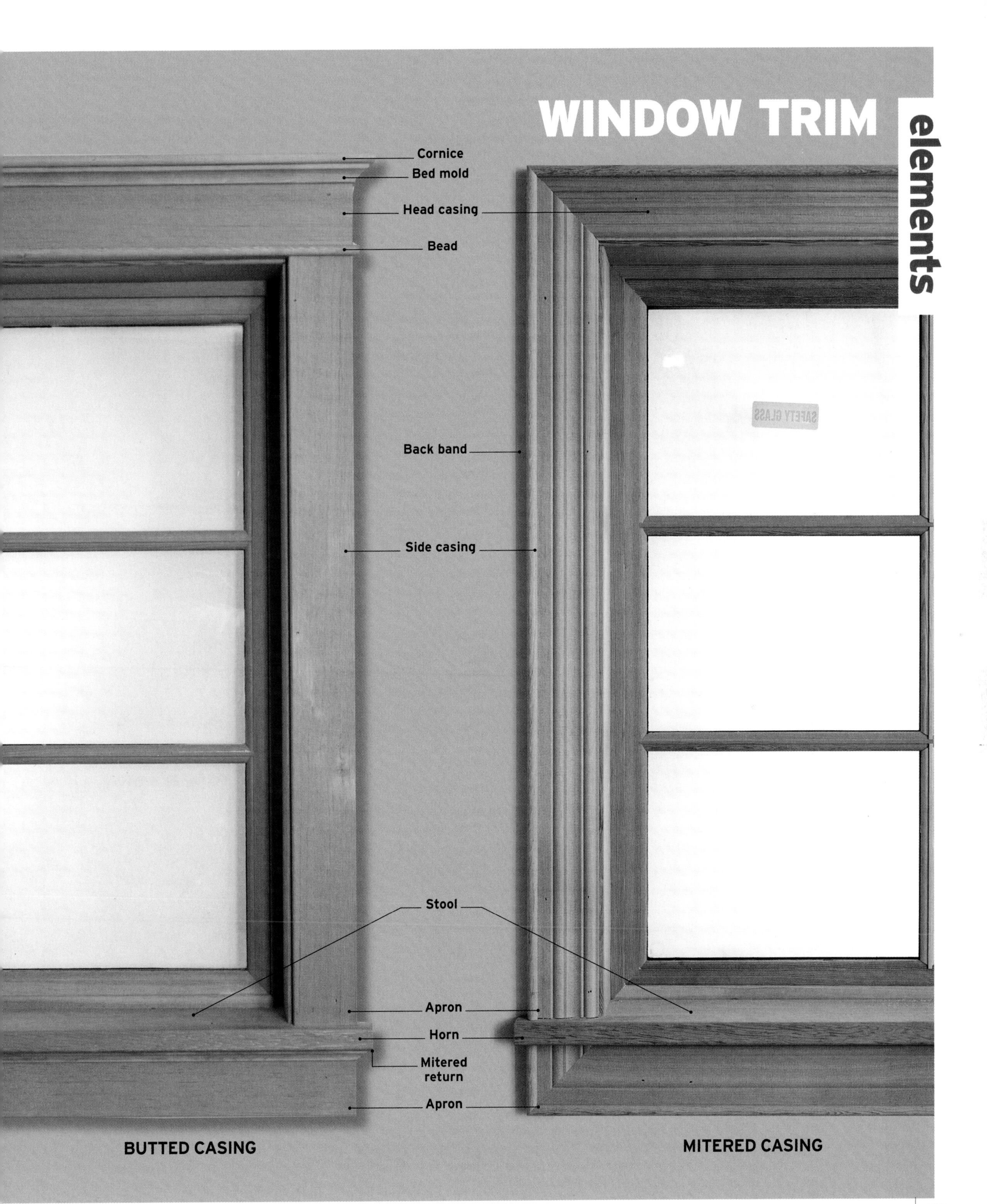

BUTTED CASING

MITERED CASING

design options

WINDOW TRIM

In a formal, period-style room, trim can be quite elaborate. A grand profiled header cap standing on dentil molding makes a dramatic statement, especially when combined with fluted mullions and neo-classical capitals. Fancy trim elements can often be found in restoration supply catalogs.

In Arts and Crafts style, trim is a dominant architectural element. Flat window casings are often wide and joined with butt joints rather than being mitered. The door and window casings are often consistent throughout the house and finished in deep natural tones.

Windows define style

The mitered surround is the most common window casing, especially in contemporary-style homes. Well-executed mitered trim presents a clean, efficient frame for a window for the window. But on wide casing trim, mitered joints may not be the best choice. They require more fussing than butt joints to achieve a tight fit. Also, if the casings shrink or expand just a fraction, a mitered joint will open up more dramatically than a butt joint—and the wider the casing, the more dramatic the gap.

The stool and apron window casing appears in most traditional styles. While most of the components of the window trim package remain the same, the profiles of the trim vary from flat (as in Craftsman) to shaped exuberantly (as in Victorian) and as such often define a particular style period. The shaped casings are either butted to rosettes or to parting beads—or occasionally mitered. Flat casings are either butted or mitered. Profiled moldings used for aprons are mitered at the ends and returned back to the wall while flat casings are usually cut off flat with the edge being slightly relieved.

Photo by Scott Gibson, courtesy *Fine Homebuilding*.

This window trim takes its inspiration from the round-overs used on the edge of the built-in storage seat, the stair treads, and the roundover trim capping the baseboard molding. The roundover header cap and stool on the window echo the theme.

CHOOSING TRIM STYLES

The first place to look for inspiration for window trim is the architectural style of the house. Colonial homes, including Capes and saltboxes, can usually take more decorative trim. Modern homes such as ranches look better with simple unadorned trim. Some modern styles look best with minimal trim. Arts and Crafts period homes have a distinctive style, whether bungalows or larger homes. Try to harmonize window casings with the other trim in the room. Since Victorian trim is so elaborate, you may need to turn to restoration catalogs to match some of the elements.

Photo by Christian Korab

When the bottom of a window jamb lacks a flat stool, as shown above, the best trim strategy often calls for miter joints on all four corners. Trim components intended for natural finish as shown should be matched for color and grain match before installation.

ADJUSTING WALLS AND JAMBS

If the drywall stands proud of (overhangs or protrudes past) the jambs (most commonly due to the window having been set crooked in the wall frame and excess joint compound build up at the drywall seams), the casing will not lay flat against both the wall and casing, throwing off joint lines. If the jamb stands proud, there will be a gap on the backside of the casing.

Drywall overhangs jambs

If the drywall stands out over 1/8 in. along the length of the jambs, ❶ it's best to apply a spacer to the jamb to build it out flush ❷. Rip and plane the spacer to the necessary thickness then pin-nail it in place, giving it a slight reveal of its own. (If it's set perfectly flush, it will invariably look a correction). Apply the casings over the spacer with their own, second reveal ❸.

Jambs overhang drywall

To deal with jambs that stand proud, the best solution is to plane them. Use a hand plane for making minor corrections ❹, but avoid using it extensively as changes in grain direction, especially across the corners, can cause tearout if you aren't vigilant about always planing in the direction of the grain. A power plane will remove more material faster and with less chance of tearout ❺. After planning or routing, even out and remove the sharp edges of the re-worked jambs with a block plane followed by a sanding block.

TRADE SECRET

To prevent scuffing the wall with the plane, apply several layers of masking tape or tape a thin shim to the rear of the plane sole.

DRYWALL OVERHANGS JAMBS

1 This drywall stands too proud to be flattened out.

2 A spacer sized to the drywall protrusion and set to a reveal.

3 Casing applied over the spacer creates an attractive, double reveal.

JAMBS OVERHANGS DRYWALL

4 Use a hand plane to trim back a slightly proud jamb.

5 Use a power plane when the excess is 1/8 in. or more.

START WITH A "JAMB" SESSION

Applying trim to a window actually starts with the installation of the jambs—the trim that fills the space in between the window frame and the face of the inside wall. Higher quality wood- and fiberglass-framed windows usually come with extension jambs already installed or cut to length and ready to be nailed on. If not, you'll have to make your own.

Jamb materials

MDF is a common choice for site-built jamb material because of its low cost and paint-ready surface, but it needs to be thoroughly painted wherever it will be exposed to moisture. It requires more shimming than solid wood to stiffen it up, especially when using it as stools. Finger-jointed pine is a good compromise for paint-grade work as it is less susceptible to water damage and it's much stiffer—though it does takes more time to finish than MDF because the joint lines between the pieces tend to show through the paint. Solid wood, though more expensive, is still the best choice for jambs and stools.

Assembly and installation

I nearly always pre-assemble the jambs using screwed butt joints to fasten the sides, top, and bottom together. Pre-assembly is much faster and more accurate than cutting and fitting each piece in place and it ensures that the joints will always remain tight. When there are more than a couple of windows to do, I find it more efficient to make a cut list and assemble all the jamb units at one time. After assembling the jamb unit, I set it in the opening on plumbed and leveled shims and nailed or screw it into place.

While jamb extensions for vinyl windows have to be nailed into the surrounding window framing, with wood windows you have the option of fastening the jambs directly to the window eliminating nailing through the face of the jambs and having to deal with shimming the side jambs.

TYPICAL WINDOW ASSEMBLY

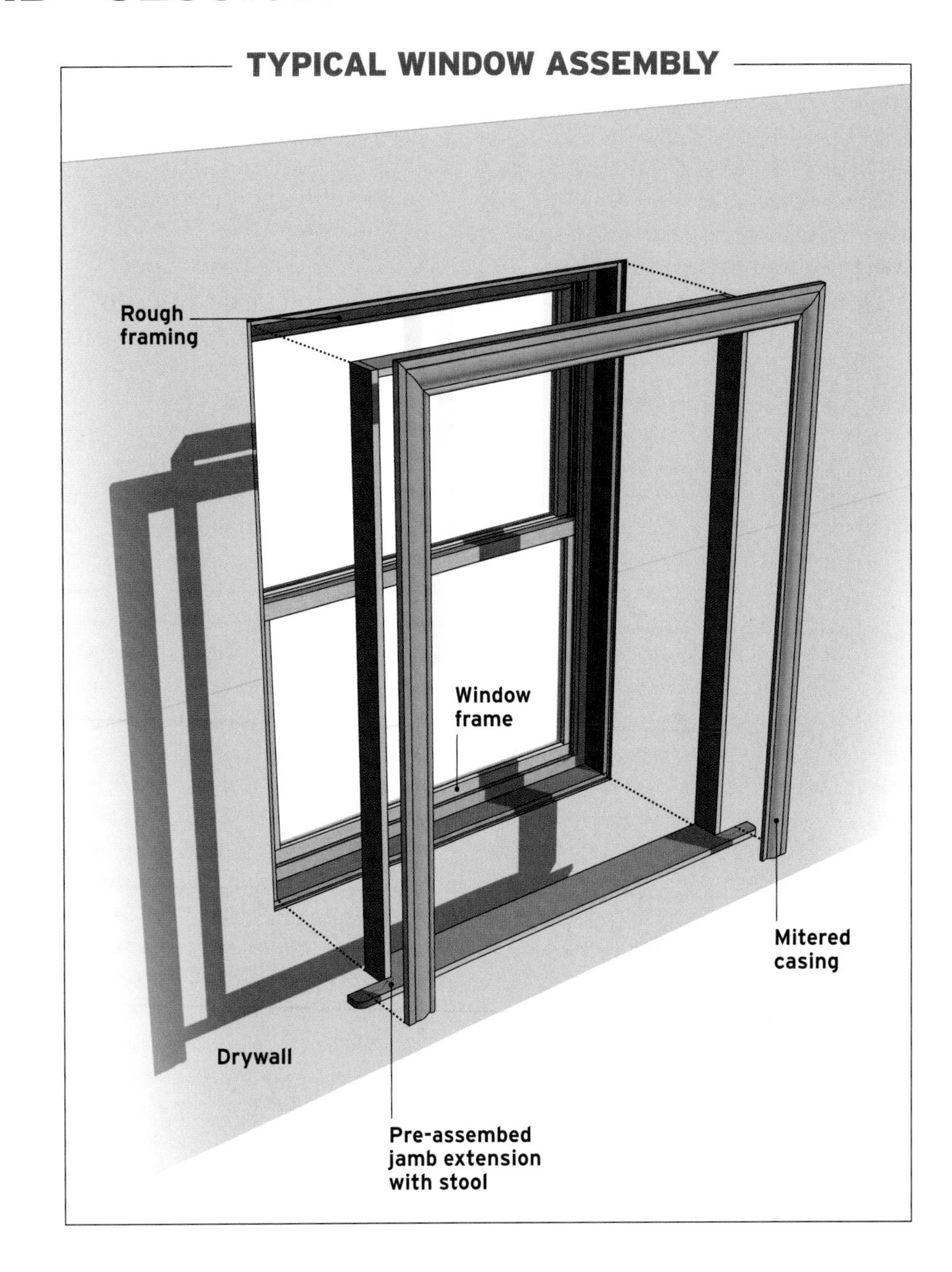

MAKING JAMB EXTENSIONS

1 Measure from window frame to wall surface at each corner for width.

2 On a flat surface align the edges and tack the pieces together.

3 Counterbore for the screws.

4 Screw the frames together with 1½ in. screws.

Start by measuring the outside dimensions of the window frame, then cut the top and bottom jamb pieces to the full length of the window frame. Figure the sides by deducting the combined thickness' of the top and bottom jambs (see the top drawing). If there is a thicker stool instead of a bottom jamb board, then be sure to include that dimension in your calculations.

To determine the width of the jambs, measure the distance from the window frame to the face of the Sheetrock at each corner and take the average dimension for all the pieces ❶. Rip all the pieces to width with a 5-degree back bevel to ensure a snug fit of the inside edge to the window frame. After the pieces are cut put them on a flat surface—bevel side up to prevent nicking. Tack-nail the two side jambs between the top and bottom jambs ❷. Be careful to keep the corners lined up and the edges flush while nailing. Now counterbore ❸ and screw the joints together using 1½ in. screws ❹.

EXTENSION JAMB DIMENSIONS

Outside dimension of jamb extension equals outside dimensions of the window frame.

Siders are full length.

Top and bottom are to outside of window frame minus 2x jamb thickness.

TRADE SECRET

Shims can either be flat or tapered and I use both when installing preassembled jambs. I make flat shims ahead of time out of various thickness of plywood and scrap vinyl and purchase commercially available tapered shims made of pine or cedar. I nail the flat shims to create level and plumb bearing points to the bottom and to one side of the framing prior to installing the jamb unit. I then use opposing tapered shims to secure and straighten the other side and top jamb and to adjust between the flat-shim bearing points.

INSTALLING SHIMMED JAMBS

It's best to keep the jambs level and plumb even if the window frame isn't—otherwise the casing trim will need much fitting to install properly.

Begin the installation by installing flat shims (made from 1/8 and 1/4-in. plywood and vinyl flooring scraps) near the corners of the framing on the bottom ❶ and along one side spaced about 2 ft. apart ❷. Size the thickness of the shims to fill the gap between the window and wall frame. You may, however, need to readjust them level and plumb ❸ relative to your carpenter's level even though that may ultimately produce an uneven reveal between the inside of the window frame and the jamb. Nail the shims in place with their front edges even with the wall surface ❹.

Now set the pre-assembled jamb unit in place on the sill shims. Keeping the jambs tight against the window frame, nail through the face of the jambs and shims into the sill and wall framing. Insert tapered shims near the corners of the other side to snug the jamb unit in place ❺. Now shim between the corners of the side and bottom jamb to make the jambs meet a straightedge spanning from corner to corner ❻. If the top jamb is over three feet long shim and check for straightness there as well.

1 Flat shims on sill create a level base for jamb assembly.

2 Nail flat shims to one side of window.

3 Check for plumb and adjust as necessary.

4 Nail the jambs through the shims into the wall framing.

5 Insert tapered shims between the wall and opposite side jamb.

6 Straighten and reinforce side using shims and a straightedge.

MAKING AND INSTALLING RABBETED JAMBS

Wood-framed windows sometimes come with rabbeted jamb extensions that fit into a matching rabbet in the window frame. Install these by predrilling holes through the jamb and then screwing or nailing them to the windows. If you don't have pre-rabbeted jambs and frames, make your own hidden attachment system by creating a tapered rabbet in the backside of the jamb stock. This creates a tight fitting, solid installation with the additional benefits of avoiding time-consuming shimming and having to face nail through the jambs.

Determine the depth of the rabbet by the length of the screw that you'll use to span between the top edge of the rabbet and into the jamb at least 1/2 in. ❶. Use the longest screw practical as the less material you need to remove from the jamb stock to create the rabbet the safer the process will be. (Note that if the jambs are less than 2 in. deep you needn't bother with a rabbet–instead predrill pilot screw holes along the face edge). See the cross section drawing on the facing page to understand how the rabbet and screw relates to the window frame.

To make the tapered rabbet, set the tablesaw to a 3-degree angle ❷. Start the rip cut 1/8-in. in from the back face of the jamb. Though this inset reduces the amount of material available to nail the casing into, the remaining 5/8-in. of the 3/4-in.-thick jamb is adequate while the inset gives the screw head more grabbing area at the base of the rabbet. Now reset the saw and make a shallow cut at the base of the rabbet to remove the waste ❸. Predrill all the screw holes ❹ and start the screws before installing ❺. Being careful to hold the jamb at an even reveal on the window frame, screw it tightly in place ❻.

1 The length of the screw determines the rabbet depth.

2 Set the blade at 3-degrees and make the first rip cut.

3 Reset the saw to remove the waste and create the rabbet shoulder.

4 Use a long drill bit to drill pilot holes for the screws.

5 Start all the screws before installing.

6 Hold jamb assembly plumb with level and screw into window frame.

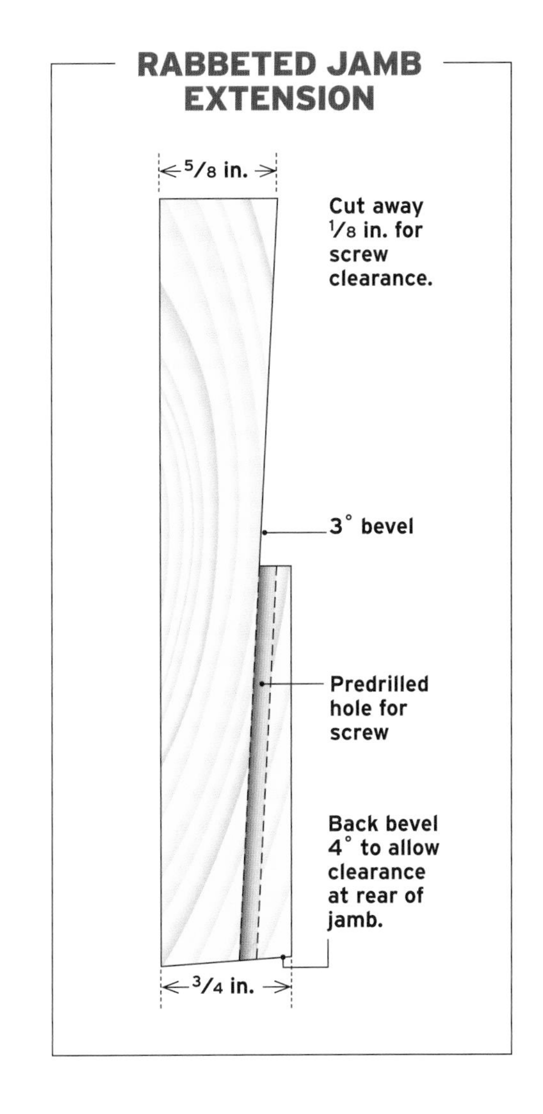

TRADE SECRET

Putting a little wax on the treads of each screw makes driving the screws easier and greatly extends the battery life of cordless drills. Use a short piece of 1/2 in. copper tubing and press the ends into a piece of toilet bowl flange wax. Dip each screw into the wax before driving. You'll find that the tubing fits handily into a work apron's pencil pocket.

MAKING A WINDOW STOOL

The starting point for making and installing this type of window trim is to create the stool. Its length and width are determined in part by the size of the casing being used, the amount of casing-to-jamb reveal and the desired overhangs past the casing. The sum of these factors is called the horn allowance. The drawing below shows how to lay out a stool and how to determine the size of the horns.

To determine the overall length of the stool, first measure the width of the outside dimension of the extension jamb unit and add the horn allowance. The horn allowance is a function of the casing width and the desired amount of reveal and overhang multiplied by two (to account for both horns). For stools that fit between sheetrocked jambs, simply add the amount of horn that extends past the window opening on each side to the length of the opening.

To determine the overall width of the stool, add together the depth of the jamb, the thickness of the casing (and perimeter molding if used) and the amount of desired overhang past the molding(s). Typically, the amount of overhang equals the amount the horn extends past the casing. For a stool fitted to a window without jamb trim, add the depth of the opening to the apron thickness and the amount of desired overhang.

After determining the overall length and width of the stool, cut the stool to finished size using a 2-degree under-bevel on the back edge. Lay out the horn length on the surface of the stool at each end ❶. Now measure the distance from the window frame to the wall surface at each end ❷ and mark this dimension on the horn layout line. Square out from this mark to the end of the stool with a combination square ❸. Cut out the waste with an inverted saber saw ❹. After the waste has been removed, use a 1/8-in. round-over router bit to form a small radius on the outside corner of the horns as well as the top and bottom edges of the stool ❺. Follow up with a sanding block to smooth the corners and edges ❻. Pre-assemble stool to the jambs with screws into counterbored holes ❼.

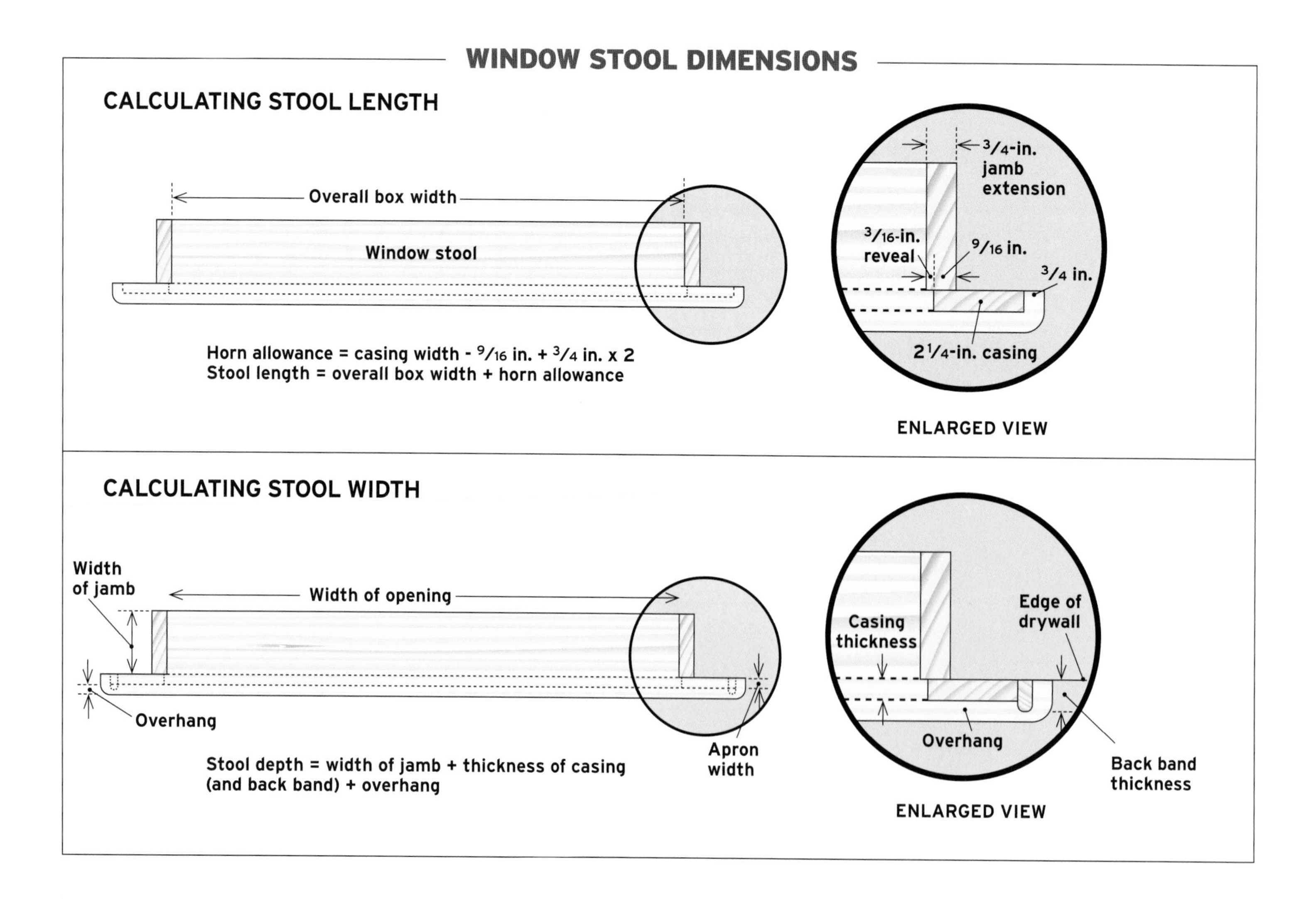

1 Draw cut lines for horn cut out using predetermined measurements.

2 Use straightedge and tape to determine width of jamb at each end.

3 Transfer distance from window frame to wall surface onto stool.

4 Cut out the waste using an inverted saber saw.

5 Round over the ends with a 1/8 in. router bit.

6 Clean up with sanding block.

7 Using counterbored holes, screw the stool to jamb assembly.

TRADE SECRET

Laying out all these dimensions full scale on a board or stick ensures that you'll produce a fail-safe cutlist since you can physically take off the dimensions rather than rely solely on calculations.

SCRIBING STOOL HORNS

1 Sticks temporarily fastened to the framing support the premarked stool.

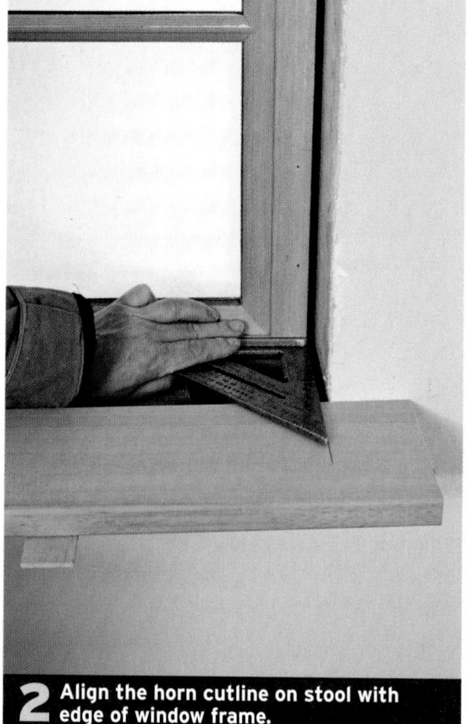

2 Align the horn cutline on stool with edge of window frame.

If the wall surfaces are not parallel to the face of the window frame, you will need to scribe the backs of the horns to ensure that the stool will fit tightly to the window frame. To scribe the cuts, hold the stool securely in place by temporarily screwing supports to the wall framing ❶. Use a square to line up the stool cutout line even with the edge of the window frame ❷. Press the stool tightly to the wall surface, then set the arms of the compass to the distance from the window frame to the back of the stool ❸. Continuing to hold the board tight to the wall, scribe the shape of the wall surface onto the board ❹. Repeat at the opposite end, then cut away the waste with an inverted saber saw.

3 Holding jamb to wall set compass from frame to back of stool.

4 Scribe wall surface to stool.

MITERED RETURN ON AN APRON

A mitered return is a fast but elegant way to finish the ends of an apron when using profiled casing. Start by cutting the apron to length with a 45-degree angle on each end ❶. Using another piece of the same material, make a 45-degree cut on each end. Now create the returns by setting the saw back to ninety degrees and, with the good-side facing down to prevent tearout, cut off the beveled portion of the miters ❷. Lay the apron on a flat surface, apply a film of yellow glue to each joint surface ❸, and then pin-nail the returns onto the ends ❹. The square-shaped, air-driven pins do not split the return and serve to clamp it in place until the glue sets up.

TRADE SECRET If you don't have a pin-nailer, substitute spring clamps or even blue masking tape to hold the returns until the glue dries.

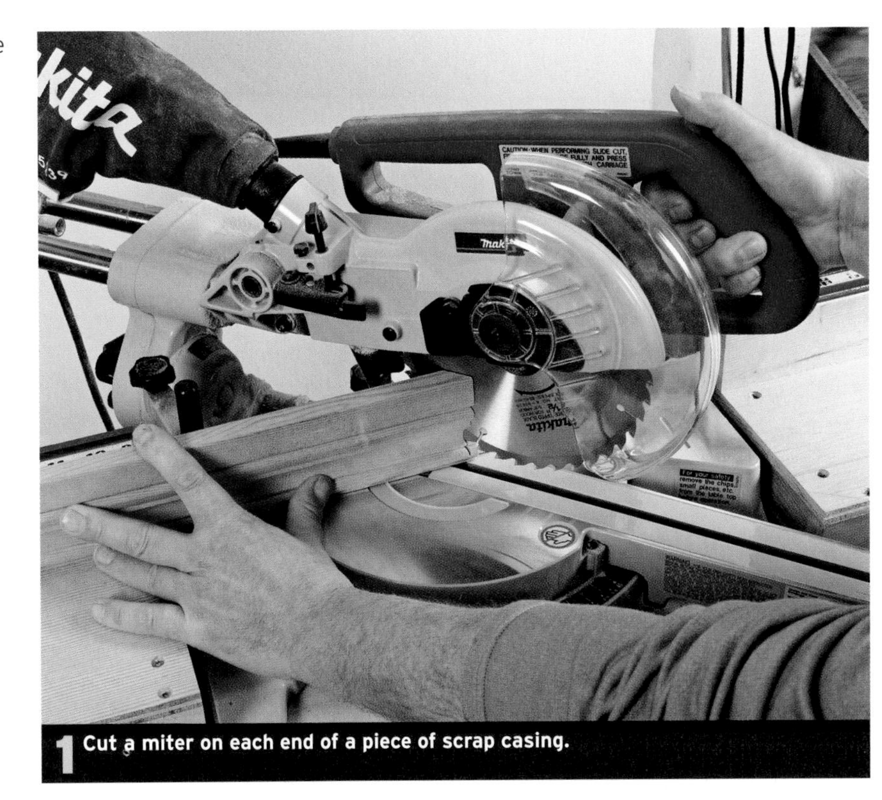

1 Cut a miter on each end of a piece of scrap casing.

2 With good side face down, cut off the beveled portion.

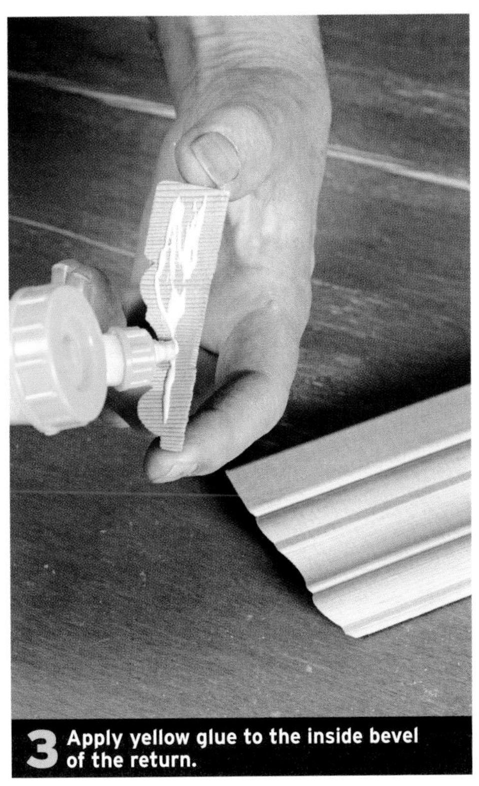

3 Apply yellow glue to the inside bevel of the return.

4 Hold the apron on a flat surface and nail in the return.

STOOL WITH DRYWALLED JAMB

Begin by holding a pair of pinch sticks between the sides of the window to record the inside width ❶. Add to this length the amount of each horn extension past the corner and cut the piece to overall length. Next, determine the angle of the right-hand sidewall to the window with an adjustable bevel gauge ❷. Draw this angle on the stool where the horn is to extend past the edge. Now hold one end of the pinch stick to the back edge of this line and mark the other end to indicate the inside cut line of the other horn. Determine the side angle at this end of the window and draw it onto the stool at the point you just marked.

Now hold the stool tight to the wall and align the end of the angled cut lines to the edges of the window opening. Use a pair of scribes to determine the opening between the window and the back of the stool board ❸. With the legs set to that distance, scribe the wall outline onto the board ❹. If the corner is radiused, use a compass to draw in the radius ❺. Now cut out the waste at either end with an inverted saber saw ❻. Undercutting the ends will allow for a tighter fit to the drywall and minimize any damage to the wall paint or the stool. Use a belt sander or small sanding drum to finish the radius before installing ❼.

Ideally, you want a fit just tight enough that the board will not quite go down all the way without a little bit of gentle coaxing ❽. If, however, it seems too tight, cut a little bit more at one end. Avoid forcing it as you run the risk of damaging the wall or the stool ❾.

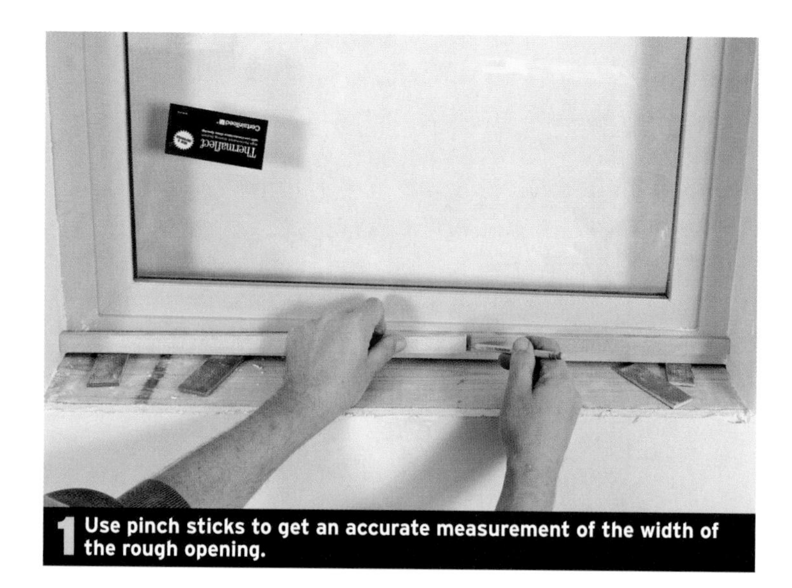

1 Use pinch sticks to get an accurate measurement of the width of the rough opening.

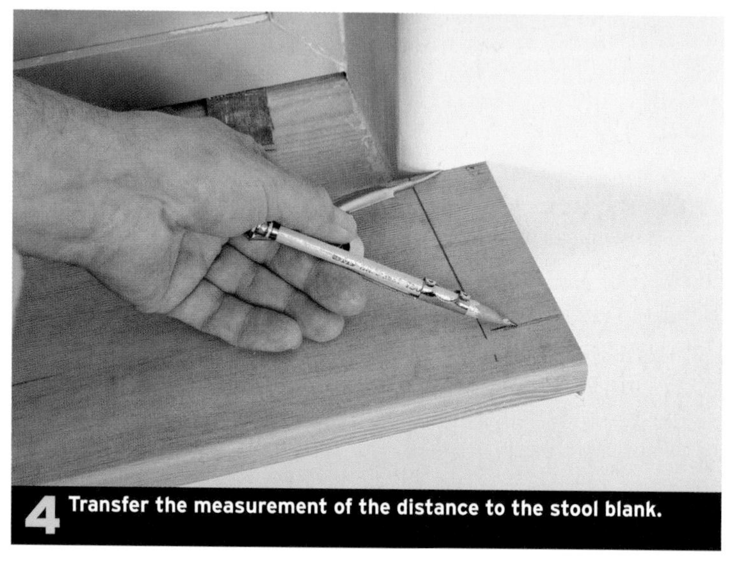

4 Transfer the measurement of the distance to the stool blank.

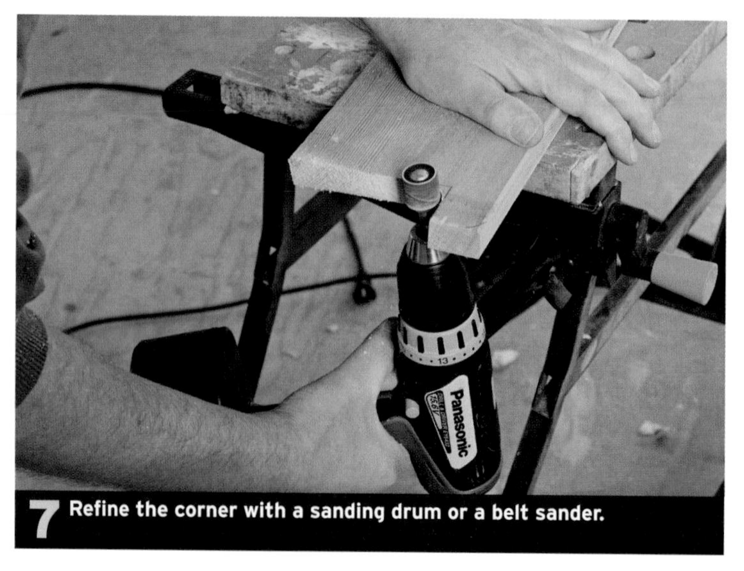

7 Refine the corner with a sanding drum or a belt sander.

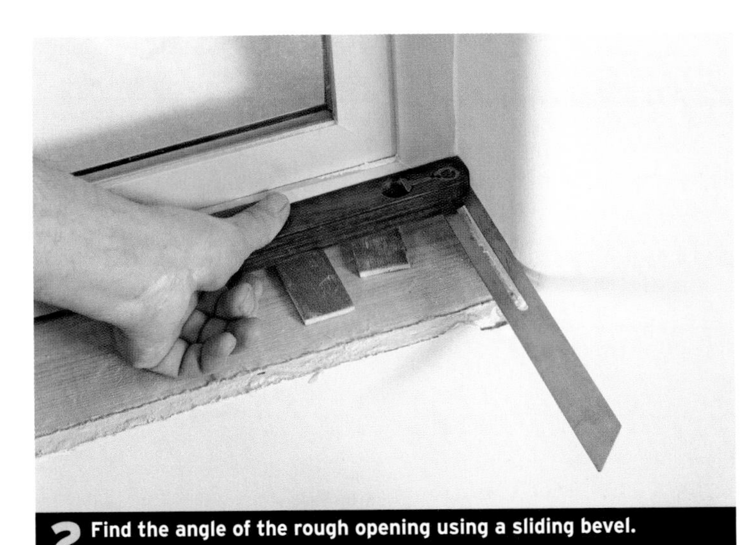
2 Find the angle of the rough opening using a sliding bevel.

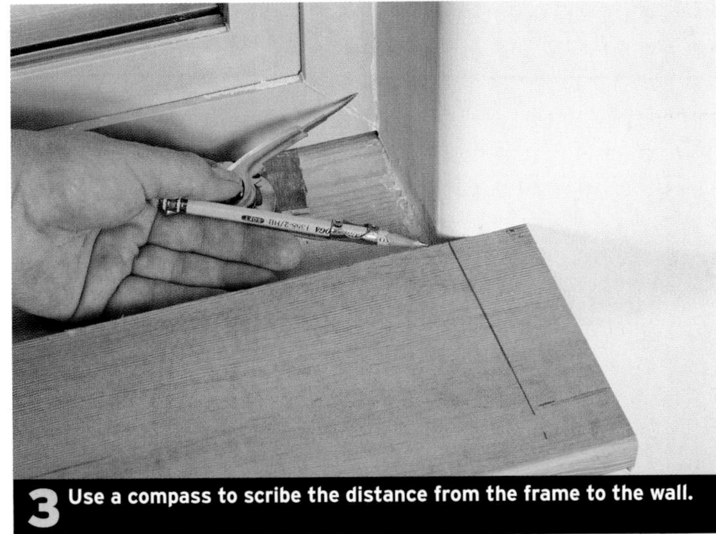
3 Use a compass to scribe the distance from the frame to the wall.

5 Use a compass to draw in corner radius.

6 Use an inverted saber saw to cut away the waste.

8 Finished stool should fit tightly into opening. Better to cut slightly oversize and trim to fit.

9 The finished stool.

MAKING MITERED WINDOW CASING

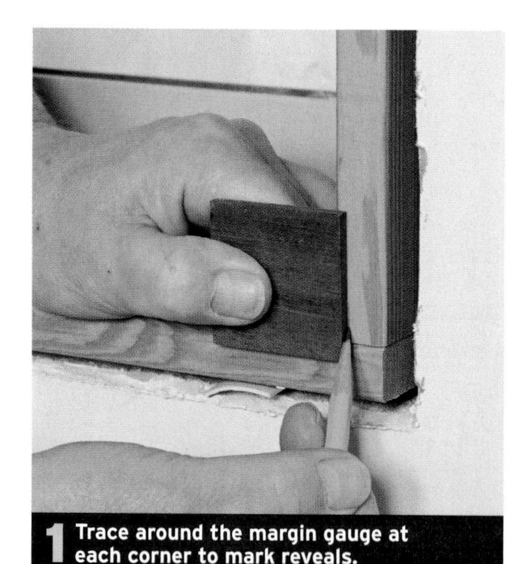

1 Trace around the margin gauge at each corner to mark reveals.

2 Mitered end of first piece installed to reveal mark.

3 Mark the opposite end for length.

4 Hold the second piece up to miter to check for fit.

5 Reinforce the miter by cross-nailing through the edges.

Though there are many strategies for creating mitered casing, I find that the method shown here tends to be faster and more accurate and error-free than most. It takes advantage of the precision cutting action of the miter-trimmer and the chop-saw to allow you to shave the casings to final fit quickly and precisely. Because the miter trimmer is easily portable, you can bring it right next to the window you're working on, saving time and effort when fitting joints.

A tape measure is useful to determine rough lengths of the individual casings (I pre-cut all the casings for each window adding 1⁄8 in. to the overall length for trimming the joints to fit)—but the most accurate and efficient way to get the final fit is to mark the pieces in place after you've fit the first joint. You'll fit the joint on one end of each casing to the previous one, then mark the piece to length using the opposite reveal mark.

While preassembling mitered window casings speeds up the installation process, it's only appropriate when all the jambs are flush to the wall surface. Otherwise, if you have to adjust the casings to fit around either a protruding jamb or wall, the joints may open up as you work on it.

➔ **See "Pre-Assembling a Mitered Casing," p. 73.**

Begin by marking the reveals on the jambs by tracing around a reveal gauge at each corner ❶. Now measure between the reveal marks, add about 1⁄8 in., and then pre-cut miters on one or both ends of each of the casing pieces to create a trim "package" for that window. If you have other windows to do, for the sake of efficiency go ahead and make packages for them now as well.

Hold the inside corner of the first piece of casing on the corner mark of the first miter ❷ and mark the other inside corner for length ❸. Cut this piece to length and nail it in place. Line the next piece up on the reveal marks and check the fit of the joint ❹. Trim if necessary, as shown at top of facing page.

THREE WAYS TO TRIM MITERS

A block plane with a sharp blade makes fine adjustments.

A chop saw quickly removes material.

A miter trimmer, although primitive, makes extremely accurate cuts.

Line up the piece again and mark the next inside corner for length and cut the miter. Brush a thin coat of glue on the joint and nail the casing permanently in place to prevent the miters from opening up as you work your way around the window ❺. Repeat this process for the third piece. If the walls and jambs aren't flush, you may need to make adjustments to get a tight-fitting joint.

➔ See "Adjusting Miters," p. 72.

The last piece is the most challenging to install, as you need to fit a miter precisely on both ends. Begin by fitting the miter at the third corner while allowing the other end to overlap the casing at the final corner. Be sure to keep the overlap even with the corner of the first installed piece as you check and correct the fit with the trimmer. Now repeat this at the final corner, letting the casing overlap the third as you check the fit. When the final corner fits, keep the same angle set on the trimmer or chopsaw and slowly shave back the last miter until the piece slips perfectly into place.

A Reveal Gauge can easily be made from scrap lumber using a tablesaw or a router with a rabbeting bit. Use at least a foot-long length of stock for safety and you'll get two gauges. Cut rabbets along one side and both edges. If you don't have the right size router bit, use a larger size. Adjust the rabbet by ripping and crosscutting equal amounts at the edge of the rabbet. Cut the gauge to a handy size.

ADJUSTING MITERS

There are three factors involved in making a mitered joint fit well: the miters must be accurately fitted to one another; the corner must be installed to a flat surface; and the stock must be uniformly the same thickness.

Usually, miter saws have detents at the most commonly used angle settings, including 45 degrees. If, however, you need to make the angle just slightly away from the detent, oftentimes you'll find the machine resists allowing you to make the setting. A quick fix to get the just-off-detent angle is to place a thin shim of cardboard, credit card or laminate sample between the saw fence and the material Ⓐ. Adjust the thickness and/or slide the shim toward or away from the blade as necessary.

If the jambs stand proud of the wall, the casings will tip backward invariably opening up an otherwise perfectly good miter joint. Sometimes all you need to do is plane a slight underbevel on the miter to regain the fit. Another option is to produce the underbevel on the miter saw by inserting a shim under the stock to lift it slightly Ⓑ. Finally, if you find the underbevel isn't enough to close the joint, you can shim behind the casing to tilt it forward and close up the joint Ⓒ.

If the face of the joint isn't flat, it's likely the stock is not uniform in thickness. With flat stock, a random-orbit sander quickly levels the joint. This is not an option on pre-finished or profiled material however. In this case use a block plane to remove material from the back of the casing before it is installed Ⓓ.

FOUR WAYS TO ADJUST MITERS

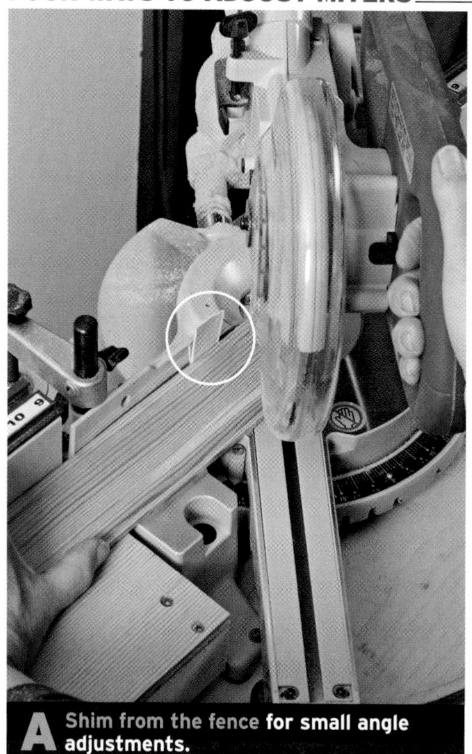

A Shim from the fence **for small angle adjustments.**

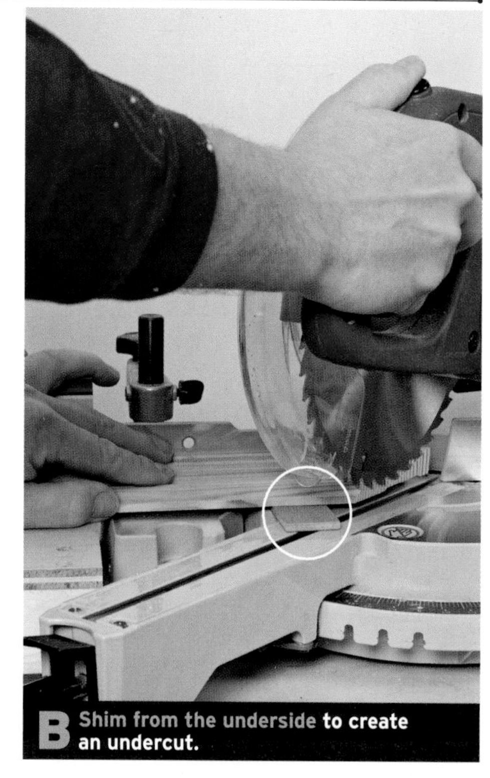

B Shim from the underside **to create an undercut.**

C Insert a shim **between casing and wall to close up mitered joint.**

D A block plane **removes material from the back of the casing to flush out the joint.**

PRE-ASSEMBLING A MITERED CASING

Pre-assembling mitered window casings can greatly speed up the installation process. Be aware, however, that pre-assembly only works well when all the jambs are flush to the wall surface.

➔ See "Adjusting Walls and Jambs," p. 58.

The key to making a successful pre-assembled miter casing is accuracy. It pays to check the accuracy of your miter saw before making cuts by fitting two test cuts around a reliable square. Adjust as necessary. Begin by cutting the opposing pieces of casing to exactly the same length on the miter saw ❶. Otherwise, the miter joints will likely not fit—or will likely open up when you try to rack the unit to fit evenly to the reveal marks on the squared corners of the jamb.

➔ See "Adjusting Miters," p. 72.

Lay out the four parts of the frame on a reliably flat surface. Glue the first miter together and clamp it ❷. Working your way around the frame, glue up the remaining miters, using clamps to hold the joints together until the glue sets up ❸. Secure the corners with brads or pneumatic pin-nails. Predrill the casing for the brads to avoid splitting ❹. Secure the casing to the jambs making sure the reveal is even around the jamb ❺.

WHAT CAN GO WRONG

If the all the jambs aren't flush to the wall, you will have to adjust the casings to fit around either a protruding jamb or wall. The joints may open up as you work on it. In addition, the jambs must be square and you must build the assembly perfectly square as well.

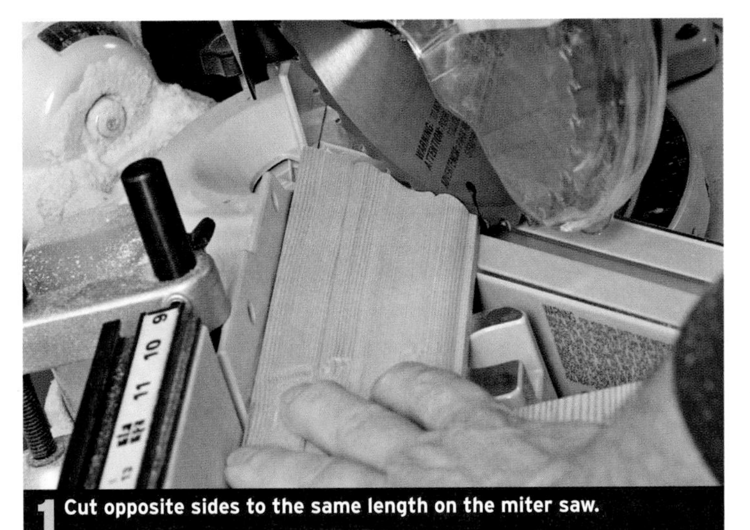

1 Cut opposite sides to the same length on the miter saw.

2 Glue and clamp the first miter together.

3 Continue to glue and clamp around rest of frame.

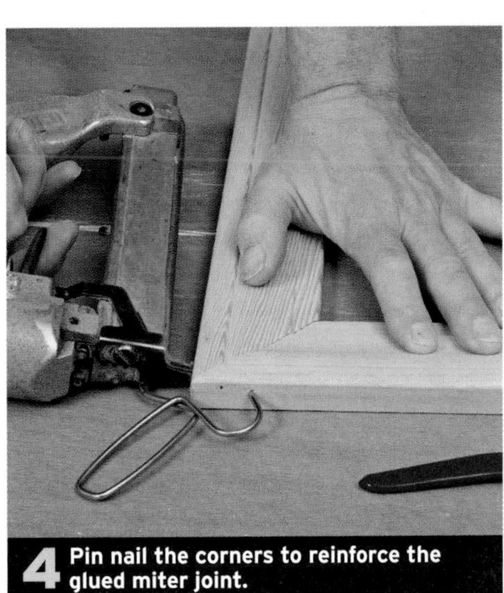

4 Pin nail the corners to reinforce the glued miter joint.

5 Install pre-assembled casings to the reveal lines on the window jambs.

STOOL FOR WINDOWS IN A SERIES

To install a common stool, you will have to cut a notch in it to clear the wall surface between the windows. You'll then level the stool on the sill framing of all the windows in a row to provide a uniform surface for all the jambs to sit on. Of course, any misalignment of the windows relative to one another may require you to compromise the margin lines of the jambs and stools with their individual window frames in order to align them level and parallel to one another. The goal is to create regular, even reveals between the casings and the edges of the jambs—and this often takes quite a bit of fiddling with the jamb installation to make it so. (If the jambs have been pre-installed and the window alignment is off, you will have to do your best to average them to minimize the unevenness of the casing margin lines.)

The only difference between making stools for windows in series and a single window is the way you measure the overall length and level and straighten the stool. Assuming you are working alone, temporarily screw a support to the sill wall frame to hold the stool in place for marking. To determine the length of the stool, set the stool on the support and align a Speed Square to the outside edge of an end window ❶. Temporarily lock it in position with a nail as shown in the photo. Line another square up at the opposite end of the windows and then measure the distance between the two legs of the squares ❷. Add this measurement to the horn allowances to get the overall length of the stool. To determine the overall width, measure from the window frame to the wall surface and add the casing thickness plus the overhang amount.

➔ **See "Making a Window Stool," p. 64.**

Cut the stool to width and length and set it back on the support, aligning a square set to the outside edge of an end window with a mark on the stool indicating the horn allowance. With another Speed Square, continue to mark the edges of the window frames on the top face of the stool ❸. Then, being careful not to move the stool, scribe the wall face profiles at the horns and intervening walls on the stool ❹. Take the stool off the support, cut out the waste with an inverted saber saw ❺ and then screw the stool to the jambs ❻. After installing shims plumb between the window and wall frame, tilt the assembly in place ❼. Level the stool by driving shims under it ❽ and then secure the entire unit in place by nailing through the side jambs into the wall framing at the plumbed shims ❾.

1 Align and secure a square with outside edge of window frame.

4 Scribe wall face profiles on stool.

7 Tip assembled unit into place.

2 Install second square and measure from edge to edge.

3 With square, draw out cut lines from the edges of window frames.

5 Remove waste using inverted saber saw.

6 Screw stool to jamb assemblies.

8 Level and straighten stool with shims.

9 Attach jambs by nailing through side jambs and shims into wall framing.

INSTALLING MITERED MULLIONS

1 Transfer reveal marks from jambs to casing to determine double-miter width.

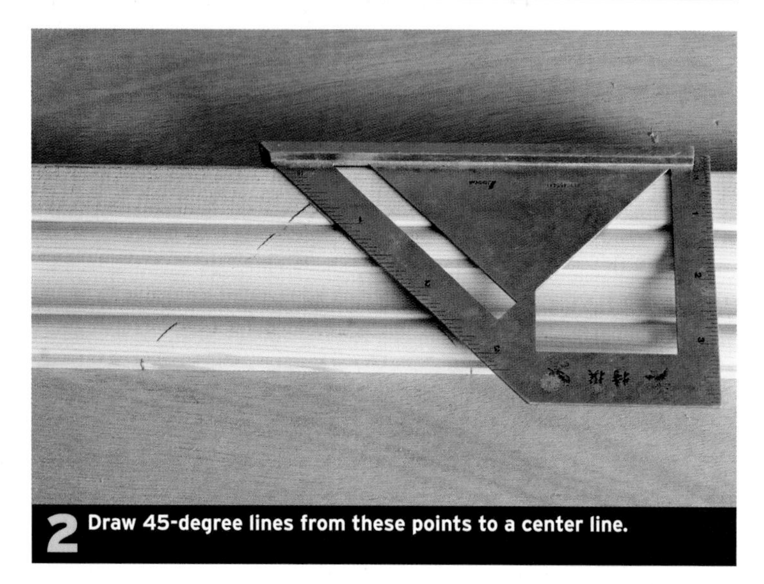
2 Draw 45-degree lines from these points to a center line.

Mullions trim out the separation between two window frames, usually covering a short span of Sheetrock. To provide a continuous profile between the casing and mullion, choose to install a mitered profiled mullion. It's not as hard to do as it looks—and it looks far more elegant than a flat molding. Begin by fitting all the casings around the window and permanently nailing the side casings in place. Cut, fit and temporarily tack the top and bottom casings in position—you'll need to remove them to cut the double-miter joint.

To make the double-miter joint, transfer the reveal marks from the side jambs onto the top and bottom casings ❶. Next, draw 45-degree lines from these marks to create a point at what will become the center line of the mullion profile ❷. Cut to the line with a miter saw set at 45 degrees, stopping each time at the point ❸. Finish the cuts with a handsaw ❹. Now reset and nail the top and bottom casings in place ❺.

Make the mullion by ripping two pieces of profiled casing to exactly ½ the width of the double-miter cut in the horizontal casings ❻. With a pinch stick outfitted with 45-degree beveled ends, get the overall length between the midpoints of the double miters ❼ and then cut the first half of the mullion to this length. Test fit, then cut the second to the same length ❽. Apply glue to the miter joint surfaces and nail the first half in place ❾. Next, apply glue along the meeting edges and miter surfaces and install the second piece. Nail along the edges of the mullion into the jambs ❿. The installed mullion gives the window an elegant finished look ⓫.

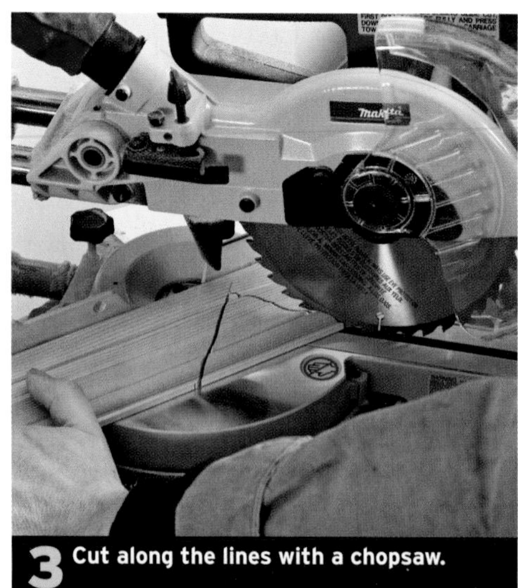
3 Cut along the lines with a chopsaw.

4 Finish up the cut to the centerpoint with a handsaw.

5 Install top and bottom casings after cutting out waste.

6 Rip two equally sized pieces to fit double-miter cut.

Clamps tighten the gap If the pieces resist closing up tightly, set a clamp across the jambs to compress the mullions together. While the clamps are still on, drive additional nails through the jambs into the window frame to help secure it permanently in this clamped position.

7 Measure length between center points with a pinch stick.

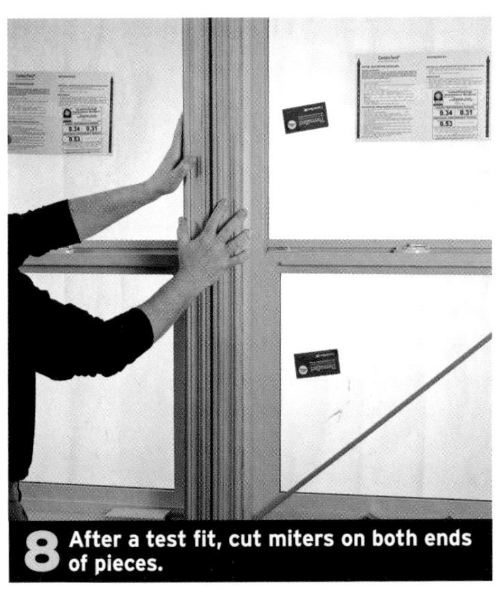
8 After a test fit, cut miters on both ends of pieces.

9 Install first half.

10 Install second half.

11 Finished double-mitered mullion in place.

INSTALLING FLAT MULLIONS

A flat mullion between windows in a series complements both profiled and simple flat casing. Measure between the reveal marks on the side jambs to establish the mullion's width ❶. If the width changes from top to bottom, you may need to cut the mullion at a taper.

Plane the stock to the correct thickness. When planning down thin stock in a thickness planer, prevent snipping and potential cracking by installing a temporary platform on the planer bed ❷. The thickness of a flat mullion depends on the casing it is butting into. On most profiled casings, the stock is usually only 3⁄8 in. thick or even less to allow it to butt against an edge profile without overlapping ❸.

➔ See "Planers," p. 39.

After planning the piece to thickness and ripping it to width, cut it to length using either a tape measure or pinch stick (add 1⁄8 in. for fitting) between the top and bottom casing to get the dimension. Line the mullion up with the reveal marks and trim the bottom joint to fit. Fit the top joint by setting the fitted bottom, then run the mullion past the top casing and marking the intersection. Be sure to mark both sides to account for an angle other than 90 degrees. Cut, trim to final dimensions, and then nail the mullion to the jambs.

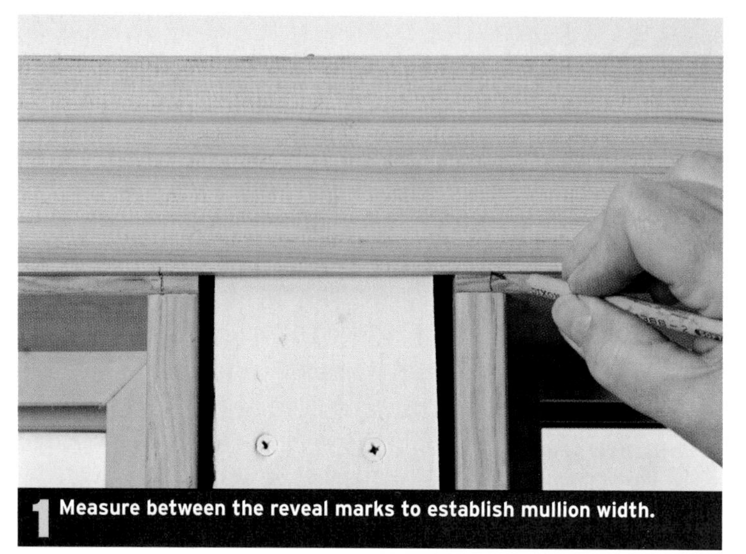

1 Measure between the reveal marks to establish mullion width.

2 Plane the mullion to correct thickness.

3 The mullion is flush with the thinnest part of the profile.

Support the back Drive in a series of drywall screws flush to provide backing. If the jambs stand proud of the drywall, you must support the back of the thin mullion trim or it may crack when pushed on. A thick bead of adhesive will often do the job but if the space is large I prefer to drive screws into the framing every 8-in. or so, setting them flush to the jambs and providing a precise and solid bearing surface.

INSTALLING CRAFTSMAN-STYLE TRIM

1 **Use the reveal gauge to mark inset of casing on the jambs.**

2 **After fitting bottom joints mark the side casings for length.**

3 **Nail the sized casings to the jambs at the reveal marks.**

Begin installation by marking the inset of the casings on the jambs with a reveal gauge ❶. Be sure to cut the side casings a bit long so that you can trim-fit the bottom joint at the stool and at the header if necessary. (Sometimes an out-of-square window frame will cause gaps.) If the bottom joints fit well, mark the length of the side casings at the reveal line at the top jamb ❷. If not, hold a level or straightedge across the top of the side casings (temporarily tacked in place) at the top jamb's reveal mark and mark the crosscut on the face of the first casing. Cut this casing to the line, nail it in place and then set the straightedge on top of it, spanning to the other side. Again align the straightedge to the reveal marks and mark the crosscut on the other casing. Cut the second side casing to length with a square cut and nail it in place ❸.

The distance between the outside of the casings, plus the overhang at either end, determines the length of the apron and the head trim ❹. (Typically in Craftsman style, both apron and header are the same length as the stool.) Nail the head trim on making sure that the overhanging portions on each end are equal and that the butt joints are tight ❺. Similarly, install the apron ❻. Ripping a two- or three-degree bevel on the top of the apron will help make a seamless fit.

4 **Measure outside edges of the side casings for head casing and apron.**

5 **Nail head casing on with even overhang on each end.**

6 **Install apron keeping a tight fit to bottom of stool.**

INSTALLING ARTS AND CRAFTS-STYLE TRIM

Elements of an Arts and Crafts head casing before pre-assembly.

1 Use the reveal gauge to mark inset of casing on the jambs.

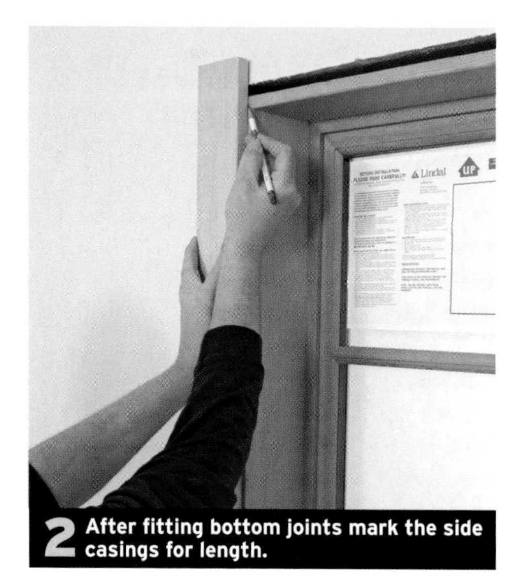

2 After fitting bottom joints mark the side casings for length.

3 Nail the sized casings to the jambs at the reveal marks.

4 Nail cove molding and returns under stool.

The Arts and Crafts style, though a bit more complicated than the Craftsman style, shares many of the some components and therefore much of the same installation sequences and techniques.

➔ See "Installing Craftsman-Style Trim," p. 79.

The only significant differences involve the addition of transition moldings—in this case a cove molding between the top head casing and its cap, a banding between the top of the side casings and the header, and a cove molding covering the joint between the apron and stool ❶. Remember to account for the presence of these moldings when calculating the length and width of the stool and the header cap.

Start the installation by marking off the reveals with the reveal gauge ❷. First measure the vertical side casings ❸, cut them to length and install them to their reveal lines ❹. Next, install the apron under the stool and apply the cove molding ❺. Miter-return the ends of the cove molding to the wall.

Pre-assemble three of the four pieces of the head casing assembly to save a lot of trips back and forth to the saw station and up and down a ladder ❻. Leave off the parting bead so you can fit it tight to the top jamb.

Then nail the bead in place to the top of each side casing ❼. Then set the pre-assembled head casing assembly on top of the parting bead and nail it to the wall framing ❽. Press the parting bead tight to the jamb while you nail it to the underside of the header casing ❾.

5 Nail the cove molding and its returns to the flat stock.

6 The finished head casing minus the bottom parting bead.

7 Nail the parting bead to top of side casings to ensure tight fit.

8 Nail the pre-assembled head casing to the wall framing.

9 Push parting bead tight to jamb and nail up into casing.

WHAT CAN GO WRONG

If the side jamb is proud of the wall surface nail the parting bead on just as it is—the gap between the bead and the wall will be mostly hidden by the side casing. If the jamb sits behind the wall surface use either a block plane or a saber saw to trim the parting bead on each end to clear the wall.

INSTALLING TRIM WITH ROSETTES

Rosettes not only add visual interest, but they simplify the joinery, hiding minor misalignments beneath their overhanging edges. Unlike mitered casings, the reveal lines are not used to determining lengths–only indicate the side-inset. Instead, you size the casings to the inside edge of the jambs where they will butt to the rosettes. Rosettes must be sized to work correctly with a particular casing width (both are sold in standardized dimensions). Matched rosettes and casings will share the same profile centerlines. You must, however, set the reveal so that the rosette extends past the casing equally on both sides.

Start installation by installing the two side casings. Fit the bottom joints tight against the stool, then mark the top cut at the bottom edge of the top jamb ❶. Cut and install to the reveal marks. Apply construction adhesive to the back of the first rosette ❷ and nail it in position tight on top of the first side casing ❸. Cut a piece of trim slightly longer than required for the top casing, align it to the reveal marks and butt it to the installed rosette. Trim to fit if necessary, reposition and mark the length to the edge of the opposite jamb. Make the cut ⅛ in. long to allow for trimming. Again hold the trim in place, place the rosette on top of the side casing and check the fit between the rosette and the top casing ❹. If it looks right, cut the top casing to exact length at the same angle–otherwise adjust the angle as necessary. Apply construction adhesive to the back of the second rosette and nail it and the top casing in place. Measure the apron to the length between the outsides of the side casings ❺. Miter cut the apron to length. For a more finished look, miter the returns. Glue a small section of mitered apron stock into the mitered end of the apron ❻. To secure the glued section, nail the return into the apron ❼. Line the apron up with the side casing and nail it in place ❽. The finished unit ❾.

1 Mark casing to length at bottom edge of top jamb.

4 Fit and mark length of head casing between rosettes.

7 Hold the apron on a flat surface and nail in the return.

2 Apply construction adhesive to back of rosette.

3 Set rosette to inside corner of jamb and nail to wall.

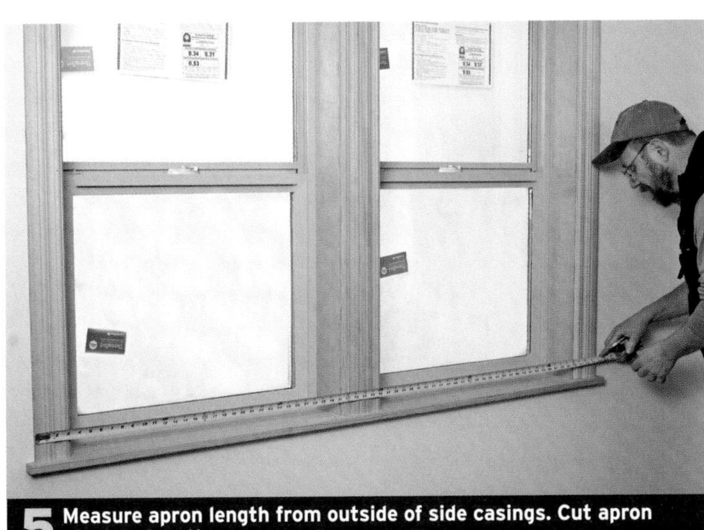

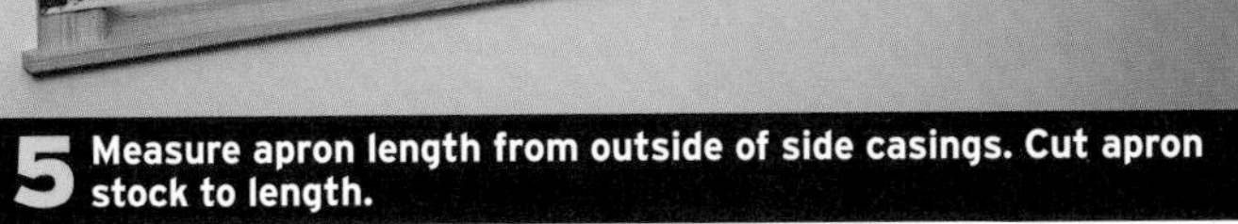

5 Measure apron length from outside of side casings. Cut apron stock to length.

6 Glue a small mitered section of the apron stock.

8 Nail the apron with pre-installed returns to wall.

9 The completed unit.

PERIMETER MOLDING AND BACK BANDING

Some traditional window trims feature back bandings or built-up moldings that run around the outside perimeter of the casings, adding a great deal of visual impact. They also add thickness to the edge of the casing, allowing wainscot or chair rail to butt into the window trim without overlapping it. These additional layers may be site-built with a simple profile or be made up of more complex purchased moldings. You install back bandings and built-up moldings after the casings are in place.

When applying a molding to the face of the casing you must be careful to keep the sides even with one another as any unevenness easily draws attention. Using a back banding in combination with an applied face molding that covers the edge joint eliminates the problem. Alternatively, a rabbeted side band also serves to hide the edge joint. Not surprising, these perimeter moldings add a good deal of time to the overall installation—though the overall effect can be well worth the effort and offer a good solution to trim intersection problems.

The trim variations shown here represent a wide variety of styles, although really elaborate perimeter moldings are usually associated with colonial styles. As you can see, back banding can be applied to casing that is joined with miters or butt joints. Most of these examples were made using off-the-shelf moldings. A few were made with custom profiles cut on a router table.

➔ **See the section on Custom Moldings, pp. 208–217.**

Install back banding in a sequence similar to flat casings. Fit and nail both sides and then fit and nail the head piece. On multilayered casings, you may need to even up the outside edges with a hand plane.

MITERED CASING WITH BACK BAND

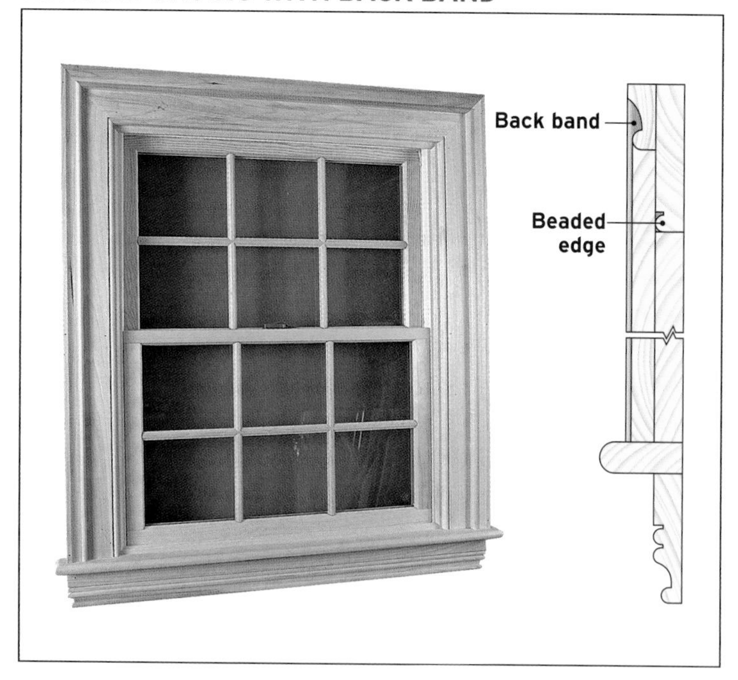

Start with a 4 ½-in.-wide flat casing and mill a simple bead on the inside edge. Run a back band around the perimeter.

INTERMEDIATE MOLDING AND BEAD

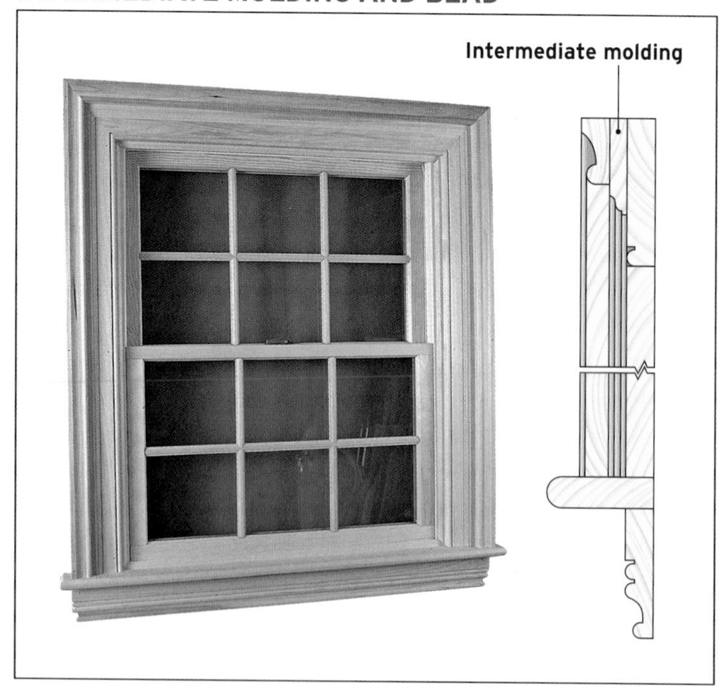

Interposed between the flat casing and the back band, a thinner (½ in. or less) intermediate molding adds another level of detail and shadowline.

SIMPLE BUTTED CASING

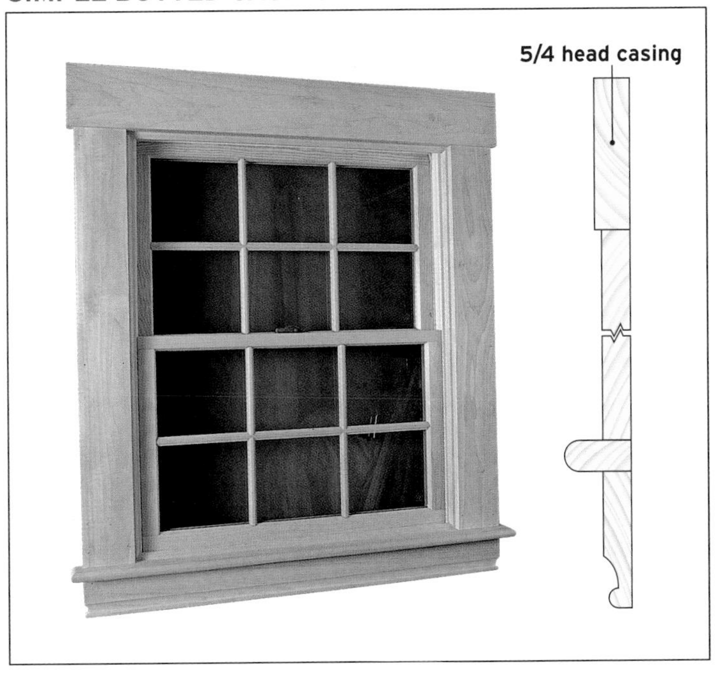

Use 4 ½-in.-wide casing for the sides, but make the head casing of thicker stock and overhang it at the ends.

LAYERS OF THICKER STOCK

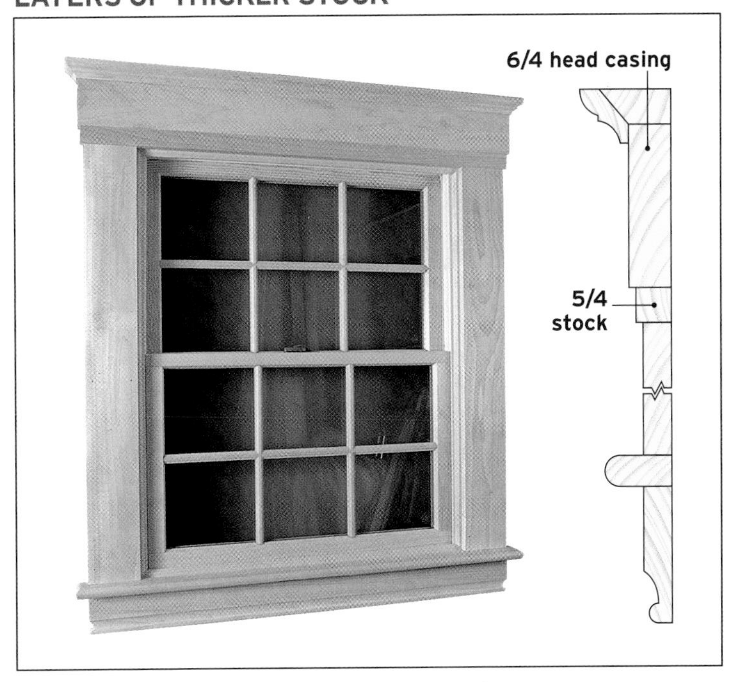

The ¾-in. side casings support a pediment made of 5/4 stock and 6/4 head casing. Cap with bed molding.

BED MOLDING AND BEAD

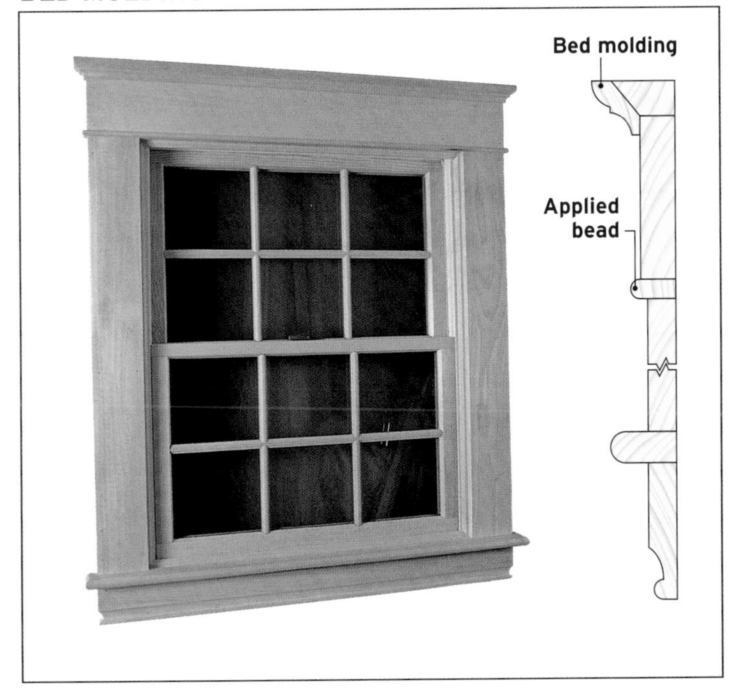

A bead below the head casing adds an elegant touch and bed molding dresses up the pediment.

SIMPLE CORNICE

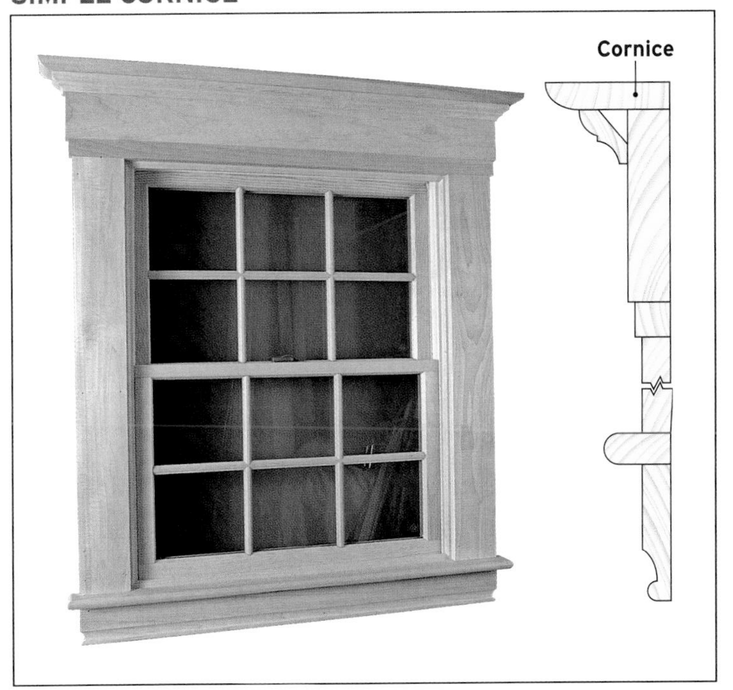

Add a simple profiled cornice above the bed mold and head casing for a more formal look.

DOORS

DOORS PROVIDE SELECTIVE SEPARATION between rooms and provide access to the house. They also express the architectural style of the home and can show off fine materials and workmanship. Doors have to function smoothly for the life of the house, and proper and careful installation is the key to years of trouble free operation. Interior doors generally come from the supplier prehung on mortised hinges in a jamb ready for installation. It's important to set the doorjambs correctly so that the door will have a solid assembly to hang from at the same time providing a flat, stable platform for nailing the trim.

The practical job of door casings is to hide the rough framing as well as strengthen the connection between the jamb and the wall framing. They also add a decorative element that emphasizes the statement of the other trim in the room. You'll see in this chapter that doors are cased very much like windows, except for the bottom element.

INSTALLATION

MAKING ADJUSTMENTS

CASING & TRIM

Photo courtesy TruStile Doors

DOORS

ORDERING DOORS

When ordering doors it is important that you are talking the same language as the person you are dealing with regarding door swing. I was taught the butt-to-butt rule: When you are backed up to the hinges you are "butted" up to the "butt hinges." If the door opens to your left side it is a left-hand door (and a right-hand door if it opens to your right). Makes perfect sense and its easy to remember. After I moved to the West Coast, however, I discovered that not every one—including the suppliers—spoke this language. To some manufacturers it depends on which side the lock is on and whether the door is opening toward you or not. Now I always make sure of which dialect door talk is being used.

This arched top door is actually a solid slab of MDF (Medium Density Fiberboard) that has been routed to imitate a traditional frame and panel design. While less expensive than solid wood, these doors are easily damaged and can be very heavy.

Photo courtesy Jeld-Wen

Choosing doors

Interior doors are available in a variety of materials and types. The least expensive, lightest, and easiest to hang is the hollow core door. These are made up of two thin sheets of plywood or hardboard glued to an interior grid work of cardboard. They come in either flat panel that may be plywood or hardboard or profiled hardboard with embossed raised panels that mimic wood. The perimeter and the area around the latch hole have solid material inserted between the plywood to provide structural reinforcement. Interior doors with a solid core of particleboard are much stronger than hollow core. They are also more soundproof and are less prone to warping.

The third common type of door is the frame and panel. The panels are either made of flat plywood or are a raised solid wood panel—both are fit into a groove cut into the frame of the door. A lot of the wooden raised paneled doors I see these days have frames made of a particleboard or MDF core with a veneer overlay. With straight-grained wood getting harder to obtain, this is obviously an economical choice on the part of the manufacturer. Veneered MDF is also quite stable, ensuring the door won't shrink or expand. The downsides are problems with the veneer lifting off. You also have to be very careful not to sand through and expose the substrate.

Interior doors usually come from the supplier pre-hung on and ready for installation. They can be either single doors or doubled to create a French door. All swing doors come with one or both edges beveled in order to help the door clear the edge of the jamb while maintaining a small margin between the door and the jamb.

Photo courtesy Milgard

French doors come in triples as well as doubles. In this door, one panel is fixed and the active door is hinged to it. The trim used here is a traditional trim with parting bead that blends well with Arts and Crafts and or prairie style architecture.

Traditional solid wood doors were frame and panel to allow the wood to move with changes in humidity. Nowadays you can get the solid wood look from composite materials covered with veneer. A mitered casing is a nice complement.

Photo by Randy O'Rourke

Glass panels known as "lights" allow light to come in from outside and visually open interior spaces to one another. These doors harmonize with the arts and crafts style of the home.

Photo courtesy Woodport

PREPARING THE ROUGH OPENING

Start by checking the rough opening for correct width and height and then mark the hinge locations on the door framing ❶. To make a plumb backing for the doorjamb, add or subtract layers of shims at the hinge points until they index your level perfectly plumb ❷. If the framing is square to the wall face use flat shims cut out of 1⁄4 in. plywood and cut up strips of scrap vinyl flooring for finer adjustments. If the framing is twisted use tapered shims to take out the twist ❸. Cut doorjambs to length by marking the length on the jamb and cutting off the excess with a circular saw run against a square saw guide ❹. With the shims in place and plumb and the doorjambs cut to the correct length, reinstall the door in the jamb.

1 Mark the hinge locations on door framing.

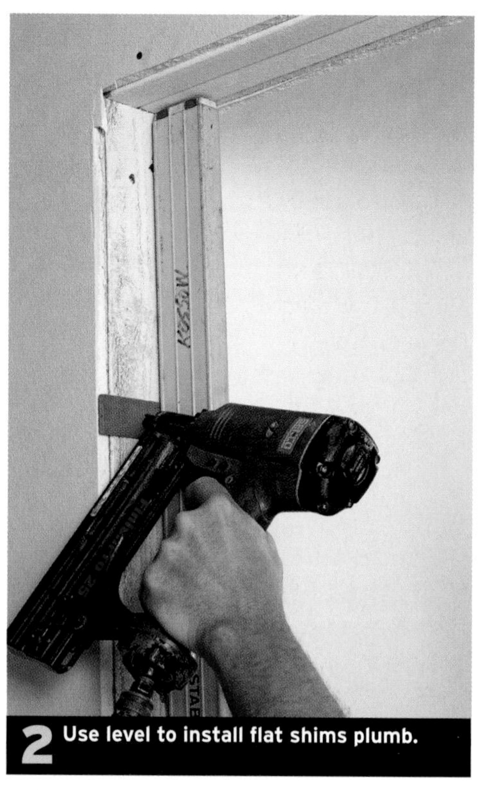

2 Use level to install flat shims plumb.

3 Use tapered shims to flatten out twisted framing.

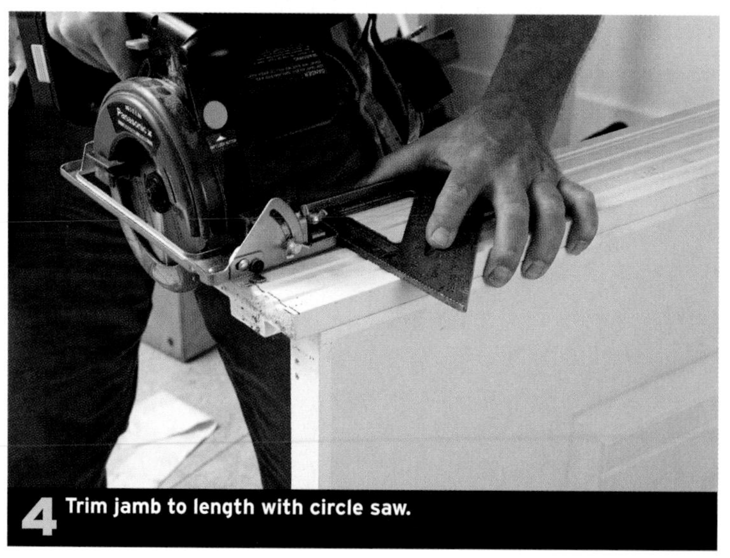

4 Trim jamb to length with circle saw.

WHAT CAN GO WRONG

Incorrectly sized rough openings are a fairly common occurrence—though thankfully it is not that much work to correct them if the opening is close to the size needed. Simply add material as necessary to one or both sides of the opening to reduce the opening. However, if you have to add so much that the casing no longer spans the gap between the jamb and the drywall, you may have to add some drywall. If the opening is too small you have a bit more work on your hands: On a non-bearing wall you could remove one or even both of the door trimmers to enlarge the opening. On a bearing wall, however, when you remove the 1½-in. studs, you should replace them with ¾-in. boards to minimize the loss of load bearing structure.

SETTING A SWING DOOR

Setting a prehung swing door is essentially a matter of plumbing the hinge side and then setting the gap margins evenly around the door. I usually install the jambs with the doors attached, though I may remove them if their size and weight are too hard to handle. Once the hinge side jamb has been set, however, I always reinstall the door before continuing with the rest of the installation.

Unless the rough opening is too narrow to allow room for shims, I always start the installation by adding flat (not tapered) shims to the top and bottom hinge locations until I create a plumb line between them. The shims will hold the jamb away from a warped stud as well as anything sticking out of the framing such as joint compound or partially set nails. It also ensures that I will get a true plumb reading with my level allowing me to set the hinge jamb dead flat and straight and therefore prevent hinge bind.

A properly hung and installed door will show even margins between the jamb and the door on all three sides and, because it closes evenly against the length of the stop, it will shut without any rattling. Hung plumb relative to the face of the wall, the door will not tend to open or shut on its own. Hung plumb relative to its side edge, the casings will also run plumb.

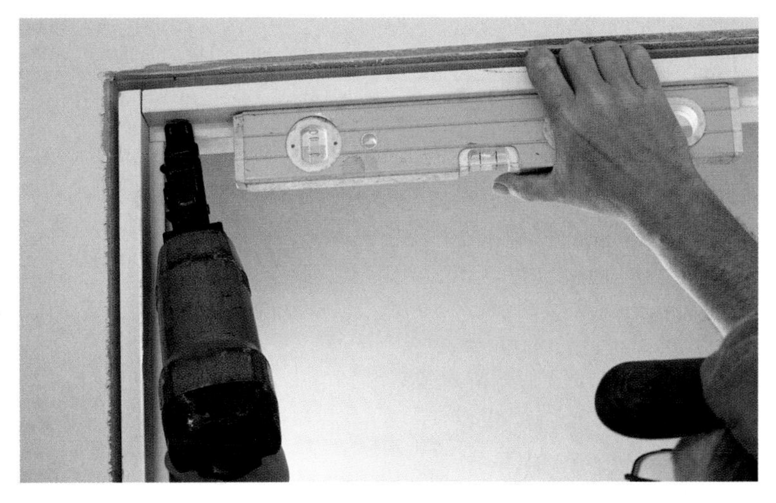

Once the hinge side is set, it's just a matter of setting even gap margins around the door.

WHAT CAN GO WRONG Sometimes, however, you may need to install a door out-of-plumb relative to any adjacent elements. When an edged-plumb door is located next to an out-of-plumb wall the gap between them appears as an unsightly taper from top to bottom. You can minimize the taper by setting the door a little out-of-plumb to match the lean of the wall more closely.

CROSS-LEGGED JAMB

When the wall framing on both sides of a door's opening are not in the same plane in relation to the wall surface the resulting situation is called a cross-legged opening. Because cross-legs prevent you from installing the edges of the doorjambs flush to the wall surface, installing the door's casing trim becomes difficult, especially at the joints. The door will also not be able to hit the stops evenly creating a rattle while shutting. While a little misalignment is workable, past a point it must be corrected. The best solution is to move the wall plates on either side of the opening. Unless there is tile or hardwood flooring run tight to the wall there is usually a little room to play with. I like driving a 16-penny nail at an angle through the plate and into the floor, which will usually move the plate. A screw also works but not as effectively as hitting the plate with a heavy hammer.

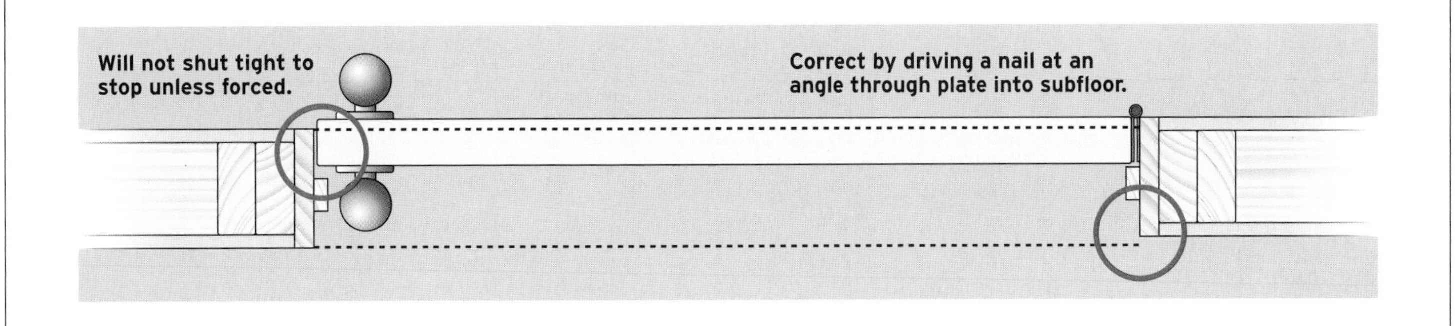

INSTALLING A PREHUNG SINGLE DOOR

1 Align edge of jamb to wall surface and nail to framing.

2 Level the head jamb and tack in position to framing.

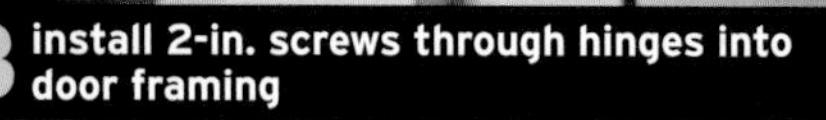

3 install 2-in. screws through hinges into door framing

or

Alternately install screws behind hinge leaf into framing.

5 Shim and tack the bottom of the jamb to the reveal.

6 Use a small pry bar under doorjamb to adjust jamb height.

7 Insert a screw through the strike mortise to secure the jamb.

After preparing the rough opening, set the door unit up, placing the bottom of the jambs on shims as required to establish the correct bottom door height. Tip the unit up and push the hinge jamb tight to the preinstalled, plumbed and flat shims. Holding on to the doorjamb, let the door swing open and align the edge of the hinge jamb flush to the face of the wall. Now shoot a nail through the jamb and upper hinge shim into the framing followed by another nail next to the lower hinge ❶. Level the head jamb in position and put one nail into the top of the opposite jamb to hold it in place temporarily ❷. Now remove an inner hinge screw from the top and bottom hinges and replace it with at least a 2-in. screw run into the trimmer stud ❸. Alternately, peel back the hinge leaf and insert the screw behind the hinge ❹. Now insert tapered shims near the top of each side jamb and adjust the reveals from side to side by the top of the door. Be aware that the tighter you shim the strike side the more the hinge side reveals gap opens up.

Now close the door and shim the bottom of the strike side jamb to create an even reveal along the edge of the top jamb. When satisfied, open the door and hold the striker-side jamb in position by shooting in one nail through a shim and into the framing ❺. Close the door and double check the reveal at the head jamb. If it needs adjusting use a small pry bar to move the strike jamb up or down ❻. Once the door shuts with equal reveals at the top and bottom of the side jambs, adjust the reveal at the location of the latch mortise and insert a screw through the mortise into the trimmer ❼. Now make sure the door shuts flat against the stops. Close the door and look where the door face hits the stops. If it doesn't hit against all the stops evenly, either the door or wall frame is warped, requiring you to shift the jambs in the opening. To do so, tap the top or bottom of one or both jamb sides with a hammer and a protective piece of scrap until the door closes evenly against all the stops. Now shim and nail behind the middle hinge and add two more nails at each shim location. Use a knife or backsaw to trim excess shim material ❽. The finished door ❾.

WHAT CAN GO WRONG

With passage doors located at the end of a hallway, it's sometimes necessary to make an adjustment so that the door and the surrounding trim look evenly spaced. The door must be centered in the middle of the end wall. In that case, you may find it necessary to alter the framing.

8 Trim any excess shim material with backsaw.

9 Properly installed, the door should swing freely.

INSTALLING PREHUNG FRENCH DOORS

1 Remove the doors from the jamb assembly before attempting to set jambs.

2 Nail jambs to preplumbed side of rough opening next to hinges.

3 Shim and drive one nail next to hinge on opposite jamb leg.

4 Use spacer to set the bottom of the jamb leg before nailing.

5 Use level and insert shim until top jamb is straight and level.

Start by preparing the rough opening. Then remove the doors from the jamb assembly and set aside ❶. Set the jamb assembly in the opening and put one nail into the prepared framing side of the jamb next to the top and bottom hinge ❷. Now shim the opposite side and shoot one nail next to the top hinge of the opposite jamb temporarily keeping that jamb leg in position ❸. Shim and tack the bottom of the second jamb leg using a piece of scrap cut to the same length as the distance between the jambs legs at the top for a spacer ❹. Now set the top jamb by using a level and inserting a shim behind the catch mortise ❺. If necessary the temporarily nailed jamb legs can be moved up or down slightly until the top jamb is both level and straight. Once the top jamb is correct nail it in place through the shim ❻. Now reinstall both doors into the jambs and check for fit. Once the doors fit correctly insert screws through the top and bottom hinges and shim and nail the middle hinges ❼. Finish up by going back and putting an additional nail through the jambs at all the hinge points and checking the margins. Both doors should have even margins on all sides and their faces should line up flush to one another when closed ❽.

6 Once top jamb is correct nail in place through the latch mortise.

7 Drive 2-in. screws through the top and bottom hinges to secure jambs.

8 Doors should have even margins and their faces flush to one another.

INSTALLING A POCKET DOOR

Pocket doors are usually sold as blanks without the jamb stock and come with square edges. Unless the doors are oversize or very heavy (as in the case of solid core doors made of particleboard) I leave them in the jambs and install them as a unit in the framing.

Initially a pocket door is hung on its rollers and pushed back into the cavity as far as it will go to determine where the leading edge of the door will end up. If needed a stop is then applied to the back edge of the door to limit how far it can slide back into the cavity. The top adjusters on the roller assemblies are used to set the door's height off the floor and to plumb the door. The bottom guides are then installed to keep the door centered in the opening. After the door has been hung, the jambs are shimmed and nailed in place using the door as a guide.

1 Lift the door into position and engage the hangers onto the roller assemblies.

TRADE SECRET To hide a pocket door track and the surrounding rough framing spray the entire track cavity a flat black before installing door.

Before installing a pocket door, read the directions on the hardware package in order to familiarize yourself with the installation requirements. Once the hangers and rollers are installed hang the door in position ❶. Use the height adjusters on the hangers to plumb the door and to set the bottom to the desired height off the floor ❷. Now slide the door back into the cavity against the stop. Ideally the door should protrude 3/4 in. past the edge of the metal stud. Most times however the door will need a stop applied to the back of the door to keep it from sliding back into the cavity too far ❸. A screw inserted into the back edge of the door works fine as a stop. Start by unhooking the leading edge hanger of the door and pivot the door to give access to the back edge. Drive a 2 1/2 in. screw halfway down the back edge of the door letting it stick out approximately 1 in. ❹. Now pivot the door back into the opening (it may be necessary to drive the screw in a little more to clear the frame), reattach the hanger and slide the door back into the cavity. Adjust the door to close evenly with the face of the side jamb by opening the door all the way and turning the screw in or out with a pair of vice grips or channel lock pliers ❺. Now attach the plastic door guides to the bottom of the metal studs, centering the door in the opening ❻. It may be necessary to drill pilot holes through the metal studs for the installation screws. The door should end up centered in the opening and slide smoothly between the guides.

2 Plumb and set the door to height by using the supplied hardware kit wrench to adjust the hangers.

WHAT CAN GO WRONG Pocket doors have a well-deserved reputation of being time consuming and difficult to install. If the pocket doorframe was installed incorrectly during the framing process it may not be possible to hang the door in it properly. In the end, it will probably save you time to tear out the frame and reinstall it. If the track was installed out of level or is not straight it may not be possible for you to install the door so that it hangs plumb both shut and open. Sometimes this is not an issue if you can set the jambs to follow the edge of the door in both positions. If the track is too out of level, however, the door will roll on it's own and in that case the sheetrock should be removed and the track reset.

3 This door will need stop applied to the back edge to keep it flush to side jambs.

4 Insert a 2½ in. screw into the doors edge to serve as a depth stop.

5 Adjust depth of screw to move the doors position when open.

6 Attach bottom door guides to metal studs.

CASING POCKET DOORS

When installing the trim on a pocket door, the length of the nails used is critical. Nails that are too long will either nail the door shut or scratch the door when opened. Of the two scenarios, I prefer the door being nailed shut. If this happens the trim can be removed, the nail pulled, and the small nail hole in the door puttied. If however the nail scratches the door it may need to be replaced if the veneer has been cut through. Avoid either situation by being aware of the nail length required. Depending on the thickness of the casings being applied use nails that only go into the jamb ½ in. or so.

INSTALLING JAMBS ON A POCKET DOOR

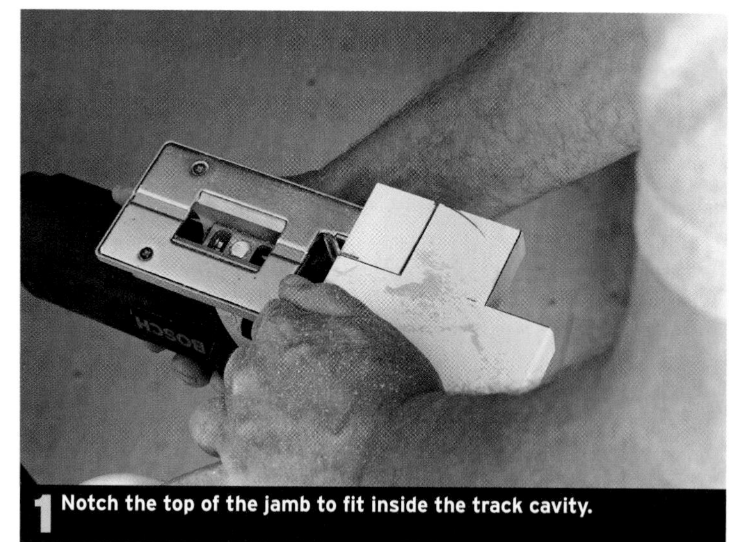

1 Notch the top of the jamb to fit inside the track cavity.

2 Shim and nail closure jamb using door edge as guide to set jamb straight.

6 Mark location of the door guide cut out on back of jamb.

7 Use repeated cuts on miter box to make clearance cut for guides.

Once you've set the doorstop and installed the bottom guides, its time to install the striker-side jamb. Notch the top of the jamb so it extends up into the track cavity ❶. Then shim and fasten the wide jamb to the framing using the closed door as a guide to indicate when it hits evenly along its length ❷.

Next install the narrow jambs on each side of the door onto the metal studs. Cut the narrow side jambs to length to fit the opening and transfer the vertical location of the rectangular holes for the nails on the metal stud onto the face of the jamb pieces ❸. Measure from the door to the center of the rectangular nail hole in the metal stud ❹. Then set the jamb in place and transfer the measurement onto the face of the narrow jambs ❺. Now mark the clearance cut out for the bottom door guides onto the edges of the side jambs using the installed guides as a reference ❻. Using a chopsaw, make repeated cuts until a clearance slot has been made for the door guides on the backside of the jamb ❼. Predrill all the nail holes at the marked locations ❽. Now apply construction adhesive at the nailing points ❾. Use 5p-finishing nails to attach the jambs to the metal stud ❿. If the side jambs do not end up even with the edge of the door, use a small pry bar to pry the jamb away from the metal stud until the surfaces are flush ⓫. The previously applied construction adhesive will lock the jamb in place once it dries. Now finish up the jambs by cutting and nailing the narrow headpieces in place ⓬.

3 Transfer the nail hole locations onto the narrow side jambs.

4 Measure distance from door face to nail cutout.

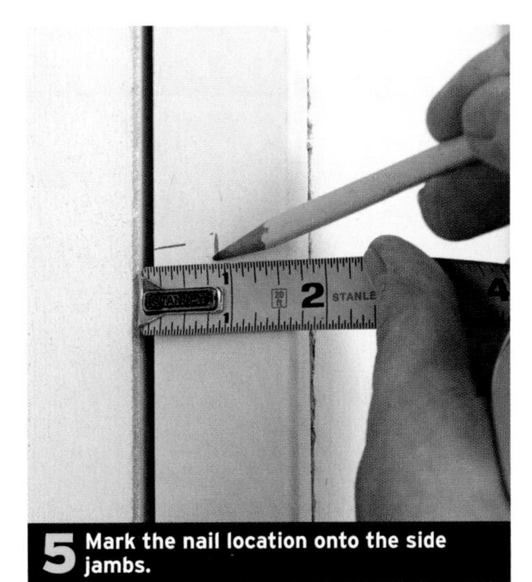

5 Mark the nail location onto the side jambs.

8 Predrill the nail holes on the side jambs.

9 Apply construction adhesive at the nailing points.

10 Nail the jamb to the frame.

11 Pry the jamb flush to edge of door.

12 Cut and nail the narrow head jambs in place.

INSTALLING BIFOLD DOORS

1 Nail the jamb assembly through the previously installed flat shim into the framing.

2 Nail the opposite jamb leg to the framing using tapered shims. Spacer board is cut to same length as head jamb.

3 Level the head jamb and nail in place.

4 Screw the track to the head jamb.

Bifold doors come sized to fit in a finished opening built to a standard door opening width. Because the doors are manufactured with the side clearances taken into account you would plan, for example, to make the distance between the jamb legs exactly five feet for a 5-ft. bifold. You'll start the installation by making sure the rough opening is large enough and then use shims to plumb one side of the rough framing before installing jambs. Assemble the jamb by cutting the pieces to length and then screwing the sides to the top.

➔ See "Preparing the Rough Opening," p. 90.

If the floor covering is going to be carpet, place a 1/2 in. spacer under the jamb feet. Nail one side of the jamb to the framing through the plumbed, preinstalled shims ❶. Since these types of doors don't hang on the jambs, nails are adequate to attach the jambs to the rough framing. Now measure or use a spacer cut to the same length as the jamb opening to position the opposite jamb leg and tack the bottom to the framing ❷. Level and straighten the head jamb and nail it in place moving the tacked jamb, if necessary ❸. Decide how far back you want the door to sit in the opening and screw the track to the head jamb ❹. If a valance board is to be used to cover the track allow room for it. Now install the bottom pivot brackets (on top of a 1/2 in. spacer block if carpet) set the same distance back from the edge of the jamb as the top track ❺. Drive the top and bottom pivot pins into the door and install the doors in place ❻. Now set the door height with the adjustable bottom pivot and set the side margins by moving the top and bottom pivot brackets along the track and bottom bracket. Once the doors are installed in place and the margins are even go back and shim and nail the side jambs to an even margin between the door edge and the jamb ❼. A properly set door will be 1/2 in. to 3/4 in. above a hard finish floor and 1 1/4 in. above the rough subfloor. After the door installation is complete, install the valance ❽.

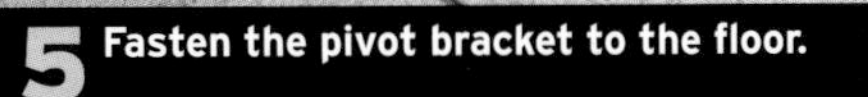

5 Fasten the pivot bracket to the floor.

6 Drive the pivot pins into top and bottom of door.

7 Shim and nail jamb using door edge to set reveals in middle section of jamb.

8 Nail the valance to the head jamb.

Bypass doors Installing the jamb set for bypass doors is similar to that of a bifold door except for determining the length of the jamb opening. Be sure to read the door manufacturer's directions: Typically the opening is 1 in. narrower than the combined width of the two doors to allow for overlap of the doors.

CUTTING DOORS TO LENGTH

1 Score door face with utility knife.

2 Align saw guide and clamp to door.

There are several things to consider when determining the bottom clearance of a door. Almost all the doors I hang come with jambs that are 3⁄4 in. longer than the door itself. When set on a 1⁄2 in. spacer (to allow for carpet) the bottom of the door ends up 1 1⁄4 in. above the underlayment.

When transitioning from a hardwood floor in a hall into a carpeted bedroom the jamb may need to be notched to drop the door down to more closely meet the hall floor. You typically want about a 1⁄2 in. between the door bottom and the top of a smooth floor covering like hardwood.

Setting doors on a solid floor can be more challenging than setting them over carpet if the floor is out of level as it will cause a tapered gap between the top of the door and the jamb. To avoid this, you'll have to shorten one of the jamb sides. On a carpeted floor you need only add or subtract some shims under the jamb legs to level the header to the top of the door.

To create a straight cut with a minimum of splintering use a shop-made straightedge guide to cut a door to length.

➔ See "Making Site-Built Cross-Cutting/Ripping Guides," p. 25.

Start by marking the door to length, measuring along both sides. If you are going to cut the door free hand to a line without the help of a guide, prevent splintering or chipping by first scoring the veneer at the cut line with a utility knife run against a straightedge ❶. If you are using the saw guide, align it to the marks and clamp it securely to the door ❷. Apply blue tape to the bottom of the saw base to prevent scratching the finish if cutting directly on the finished door surface. While keeping the saw's base tight to the guide, cut the door to length ❸. Sand the bottom edges slightly to take off the sharp edge and to eliminate any potential splintering ❹.

3 Cut door to length with guide and circle saw.

4 Sand the bottom edges of door smooth.

TRIMMING A HOLLOW CORE DOOR TO LENGTH

If you must shorten a hollow core door more than an inch ❶, be aware that the solid edging may disappear, exposing the internal cardboard grid work. If this happens another edging piece must be inserted into the cavity at the bottom of the door to reinforce it. Start by removing enough of the interior cardboard filler strip to allow room for the new edging piece ❷. Now rip a length of solid stock to the same thickness as the door cavity and cut it to the required length. Apply glue to the edges and slide the piece in place ❸. Clamp the door face veneers to the new edging strip until the glue sets up ❹. Use boards between the clamp and the door to help distribute the clamping pressure evenly and to protect the face of the door. Finish up by sanding off the sharp edges at the bottom of the door ❺.

1 Cut the door to length using cutting guide clamped to door.

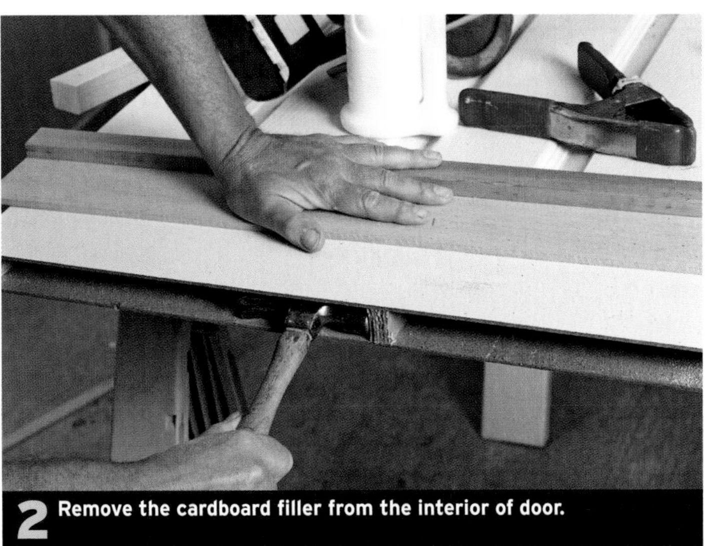

2 Remove the cardboard filler from the interior of door.

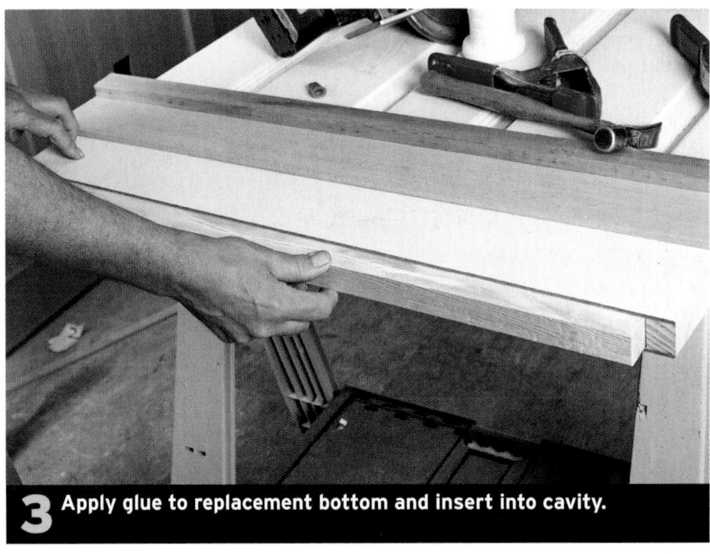

3 Apply glue to replacement bottom and insert into cavity.

4 Clamp the door sides together.

5 Sand sharp edges off bottom of door.

CUTTING DOORJAMBS TO LENGTH IN PLACE

Occasionally there arises the need to shorten doorjambs in place after they have already being installed. This can happen when a decision is made to add hardwood floor or tile in a room that was previously earmarked for carpet. While professional flooring installers often use an electric jamb saw with an offset blade designed to cut close to the floor, the result is usually a saw cut full of splintering and tearout. The best way to cut a doorjamb to length in place is to use a Japanese-style pull saw ❶. When set to a spacer cut to the same thickness as the floor, the jamb can be accurately and cleanly cut to length ❷.

1 Use piece of flooring to guide backsaw when cutting bottom of jamb.

2 Slide flooring under jamb after cutting jamb to length.

PROBLEMS WITH PREHUNG DOORS

Prehung doors are often manufactured in one plant and then sent to another where they are hung to a set of jambs. As a result, doors may occasionally arrive at the jobsite with incorrectly cut and poorly assembled jambs and improperly mortised and attached hinges.

HINGE-BOUND

If the mortises for the hinges are routed too deep the door will hit the jamb before it is fully closed creating a situation called hinge bind. (Indicated by the hinges pulling and twisting the jamb when the door is shut tight to the stop and the door tending to spring open as soon as you turn the knob.) If this is the case, put a layer of thin cardboard behind the hinges to bring them flush or a little proud of the jamb surface.

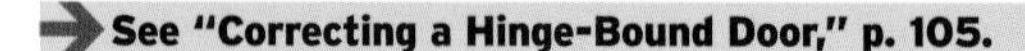

See "Correcting a Hinge-Bound Door," p. 105.

STRIPPED SCREWS

Sometimes the standard hinge screws are short and are easily stripped out. It is not uncommon to have a door arrive with several of the screws in each hinge stripped out and the hinge loose. If this is the case, the stripped-out screws will need to be replaced with larger ones or the hinge will inevitably work loose changing the doors reveals and affecting their operation.

WRONG-SIZED HEAD JAMBS

If the head jambs are cut to the wrong length it will make getting even margins impossible unless the jamb is taken apart and recut to the correct size.

WARPED

Apart from the hinge and jamb problems the doors themselves can be warped preventing the door from closing evenly against the stops. If the door is warped return the door for a replacement.

TRADE SECRET Although pass-through openings do not have a door they still require jambs and trim to cover up the rough framing. Be sure to make the jamb heights match the height of any nearby doors. When constructing a set of jambs for a hinged door the traditional way to assemble them is to insert the head jamb into a rabbet cut in both jamb legs. This is standard practice on exterior doors and on extremely heavy or oversized interior doors. Standard interior doors are simply butt jointed and stapled at the factory. On a pass-through opening you assemble the jambs on-site using butt joints and screws to hold the pieces together.

CORRECTING A HINGE-BOUND DOOR

When a door binds, the first thing to check is to see if the hinge leaves have been mortised too far below the surface of the wood on either the door or the jamb, causing the jamb and the door to touch. If a mortise is too deep cut a thin piece of solid cardboard (like the back of a writing tablet) and insert it behind the hinge to build it out slightly past the surface of the wood ❶. The second thing to check is to see if the hinge jamb is twisted toward the hinge. If it is, loosen the long screw going into the framing through the hinge to allow the jamb to straighten out and relieve the twist ❷. Nailing the opposite side of the same jamb tighter to the framing may also help pull the jamb back in line ❸. Either of these steps may require resetting the shims under the hinges. Make sure to check the door for proper margins after correcting the fit.

1 Insert cardboard shim behind the hinge.

2 Loosen long screw through hinge into framing.

3 Push jamb tightly while nailing to framing on opposite side of jamb.

PREPPING THE WALL AND DOOR FRAME

1 Check the surface alignment of doorjamb to wall surface with straightedge.

2 Flatten any protruding Sheetrock with hammer.

3 Add additional fasteners to Sheetrock if necessary.

Before installing the casings on a door or passage opening, there is often some prep work involved. The closer the wall and the face of the jamb are aligned the easier it will be to install the trim. If the misalignment in the two surfaces is too large for the particular casing being used, a gap will be created behind the casing, either at the wall surface or at the jamb. Start with a visual inspection using a piece of the casing being used in order to assess the alignment of the jamb and wall faces. Depending on the type and size of the casing either the Sheetrock or the edge of the jamb may need to be altered.

Use a straightedge to check for any misalignment between the wall and jamb ❶. Flatten any Sheetrock protruding past edge of doorjamb with a hammer ❷. Occasionally the Sheetrock will not have been fastened to the wall with enough fasteners and additional fasteners will need to be applied ❸. If the amount of Sheetrock sticking past the jamb is more than can be pounded down, remove it by scribing a line on the wall tracing the outline of the casing. Then cut through the paper with a utility knife and remove the paper and excessive drywall with a small pry bar or a putty knife ❹. If the jamb protrudes beyond the wallboard, back bevel the jamb edge with a hand plane. This eliminates any gaps by allowing the backside of the casing to lie tight to both the jamb and the wall ❺.

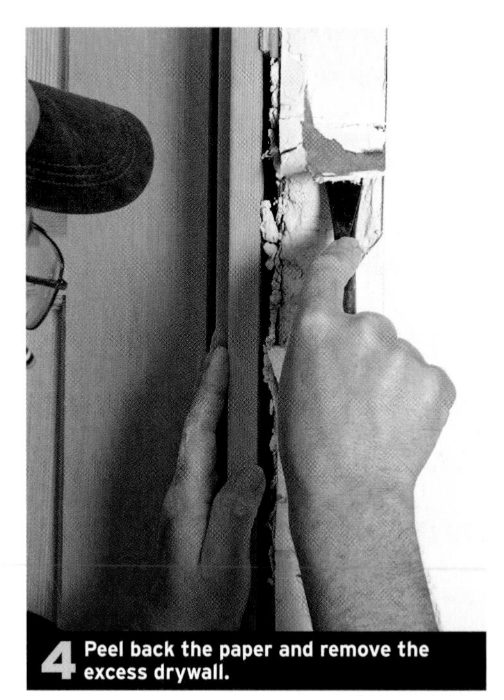
4 Peel back the paper and remove the excess drywall.

5 Back bevel the jambs to enable the casings to lie flat.

CASING WITH MITERED TRIM

Mitered door casings, like window casings, are best installed by individually fitting each mitered piece rather than by pre-assembling and installing them as a unit. I recommend reinforcing miter joints by coating the mating surfaces with glue and by nailing across the joint through the edges of the molding.

➔ **See "Adjusting Miters," p. 72.**

Precut the casings to rough length and match them for color and figure prior to installation. To install mitered door casings with a tight fit to a finished floor surface, start by marking the reveals on the jambs with a reveal gauge ❶. Then mark the casings to length to the marked reveals ❷. Cut the miter on the casing at the marked point and nail the casing to the jamb and wall. When nailing on a side casing, get in the habit of nailing from one end. This allows you to straighten out a crooked molding and create an even reveal as you nail from the fixed end toward the loose end. Now cut a miter on the end of the head casing and check the fit of the joint ❸. Once the joint fits tight with an even reveal between the casing and the jamb edge, use the reveal mark on the opposite jamb leg to mark it to length ❹. Now set the saw to 45 degrees and cut the head

1 Use margin gauge to mark reveals.

2 Mark casing to length using reveal marks.

TRADE SECRET

The process for installing door casings on bifold and bypass doors is identical to that of other doors except for the addition of one more piece of molding. Since both bifold and bypass doors utilize a metal track to hold and guide the doors, a valance is generally required to hide it. This trim is generally of the same material and finish as the door casings and jambs.

3 Check fit on first miter joint.

4 Mark head trim to length using reveal mark on opposite jamb leg.

>> >> >>

CASING WITH MITERED TRIM (CONTINUED)

5 Nail head trim in place.

6 Check fit of miter joint with short test piece.

7 Mark second casing leg to length.

8 Nail second casing leg to wall.

9 Cross nail to reinforce miter joints.

trim to length and glue and nail it in place ❺. On the second casing leg, use a test piece to get the proper fit on the miter joint prior to cutting the casing to length. Make the test piece long enough so you can line it up accurately to the reveal as you check the fit of the joint ❻. After the correct miter angle has been determined on the test piece, hold the second leg in place butted onto the finished floor and mark it for length ❼. Cut the leg to length using the angle determined on the test piece and nail it to the wall ❽. Finish up by tightening and reinforcing the miters by cross nailing the corners ❾.

TRADE SECRET When installing casings on door and pass-through openings that land on a hard finished floor (for example: tile or finished wood), the side casings require a bit more accuracy than when installing trim to carpet where there is more tolerance for irregular lengths.

CRAFTSMAN-STYLE CASING

1 Mark casing reveals with reveal gauge.

Craftsman door trim typically features a thicker head trim piece that overhangs the face and the outside edges of the side casings. Sometimes the head casing may run continuously, connecting with the head casings on the windows and forming a band that runs around the room.

Start by marking the reveal lines on the jamb edges with a reveal gauge ❶. Select only the straightest stock for side casings, especially if the style calls for trim wider than 4 in. You'll find that wider stock is difficult to straighten edge-wise as you nail it in place. Keeping the bottoms of the casings either tight to the hard finished floor or set onto a 1/2-in. spacer to allow for carpet, mark the side casings to length to the reveal marks on the head jamb ❷. Cut the casing legs to length and, while keeping the reveal spacing even, nail them in place ❸. Now determine the length of the head trim by measuring the distance between the outside edges of the casing legs and adding the combined overhang amounts ❹. After slightly sanding the edges, nail the head trim to the wall ❺.

2 Mark casing to length.

3 Nail the side casings to the jamb and wall.

4 Measure distance between outside of casing legs.

5 Nail top casing to wall and jamb.

TRADITIONAL CASING WITH PARTING BEAD

The Arts and Crafts style is similar to, though more complex, than the Craftsman. It uses the same square section casing pieces but adds a bit more ornamentation through the use of cove moldings and parting beads at the transition points. You can pre-assemble the entablature before installing it to the door as we did for the window casing.

➔ **See "Installing Craftsman-Style Trim," p. 79.**

But you can also install the trim piece, by piece. Begin by installing the side casings as in for craftsman trim.

➔ **See "Craftsman-Style Casing," p. 109.**

Determine the overall length of the parting bead by measuring to the outside edges of the side casings and adding the combined overhang amounts. Rout and sand a radius at the ends of the parting bead to match the front, then nail it in place to the top of the side casings ❶. Now nail the head trim into the wall framing ❷. Push the parting bead tight to the doorjamb and nail up into the head casing ❸. Next, install the cove molding. Mark the molding to length at both ends of the head trim by butting an end to the wall and marking the inside corner of the miter ❹. Miter-cut them to length and nail one of them in place ❺. Now cut a miter on one end of a length of cove molding and, while keeping it tight to the installed piece, mark the other end for length ❻. Cut the piece to length, apply glue to the miter joint surfaces, and nail it and the last end piece to the head casing ❼. After measuring and cutting the top cap to length nail it to the top of the head casing ❽.

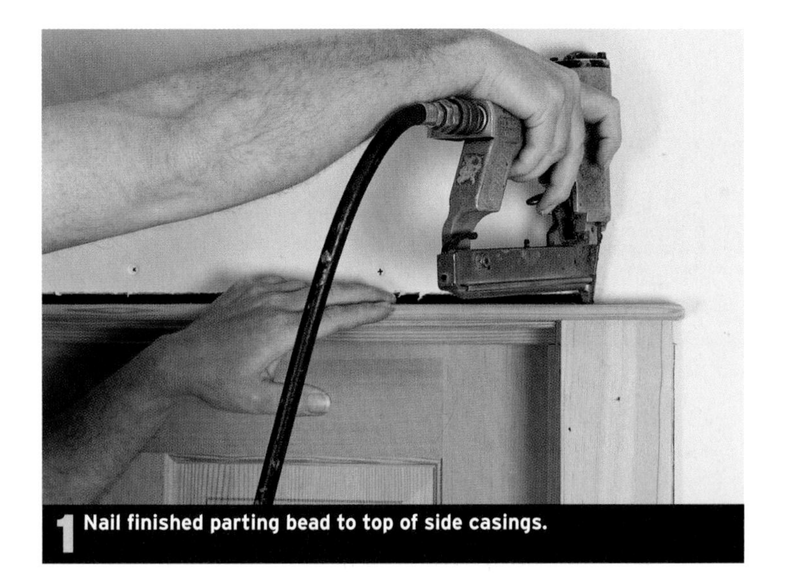

1 Nail finished parting bead to top of side casings.

4 Mark cove molding to length at both ends of head casing.

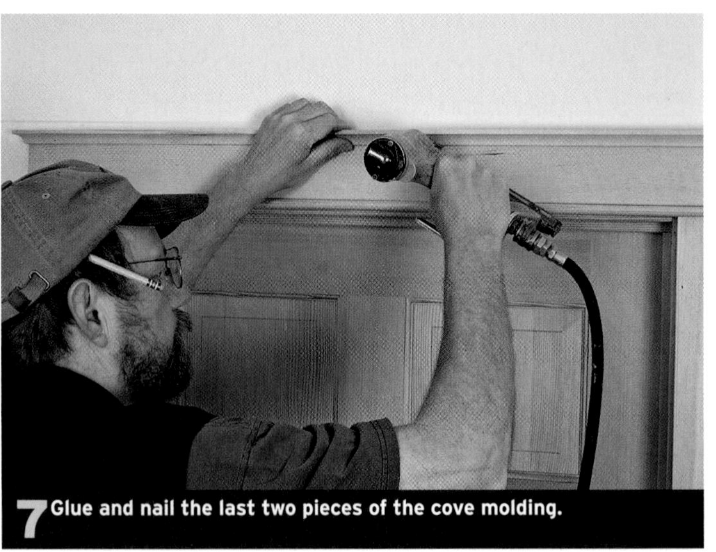

7 Glue and nail the last two pieces of the cove molding.

2 Nail head trim to wall framing.

3 Hold parting bead tight to doorjamb and nail into head casing.

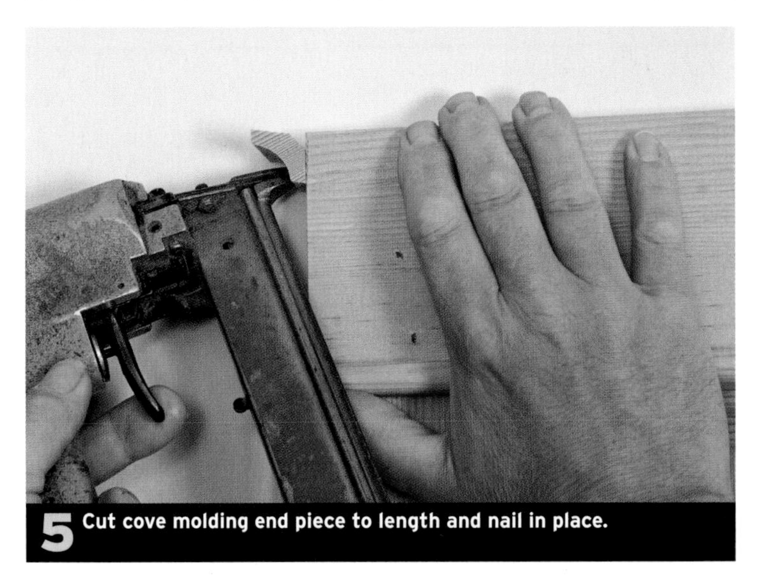

5 Cut cove molding end piece to length and nail in place.

6 Butt mitered ends together and mark opposite end for length.

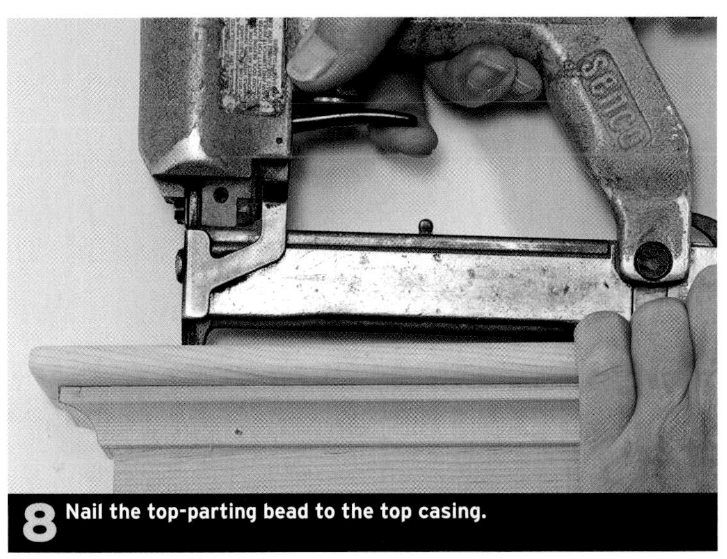

8 Nail the top-parting bead to the top casing.

CASING WITH ROSETTES AND PLINTH BLOCKS

The corner blocks and plinth blocks sometimes used in Victorian door casings add another decorative element and eliminate having to use miter joints at the molding intersections. This is a definite advantage when using the larger dimension moldings typically found in the Victorian style as it creates a joint that has less chance of opening up as the moldings go through their seasonal dimensional changes. The corner blocks are used at the casing intersections while the plinth blocks create a transition point between the standing moldings and the baseboard.

Use a reveal gauge to mark the reveals for the casings. Set the plinth blocks tight to the finished floor and nail them to the wall and jamb ❶. Cut the side casings slightly longer than required. If the joint to the plinth looks good (trim if necessary), hold the casing in place and mark for length at the bottom of the head jamb ❷. Cut the casing to length and nail it in place to the reveal marks ❸. Apply construction adhesive to the back of the first rosette and nail it in position even with the inside edges of the jambs ❹. Cut a piece of trim slightly longer than required for the top casing and check the fit of the joint while aligned to the reveal marks and butted to the installed rosette ❺. Trim this joint to fit if necessary. Now holding the casing on its reveal marks and butted to the installed rosette, mark it to length while checking the fit of the joint at the second rosette setting on top of the casing ❻. After fitting the joint and cutting the piece to length, nail the casing and rosette in place applying construction adhesive to the back of the second rosette ❼.

1 Nail plinth block to jamb and door framing.

2 Mark casing for length at bottom of head jamb.

4 Nail the rosette tight to top of side casing.

5 Check fit of top casing to first rosette.

7 Nail second rosette and top casing into place.

3 Nail casing in place.

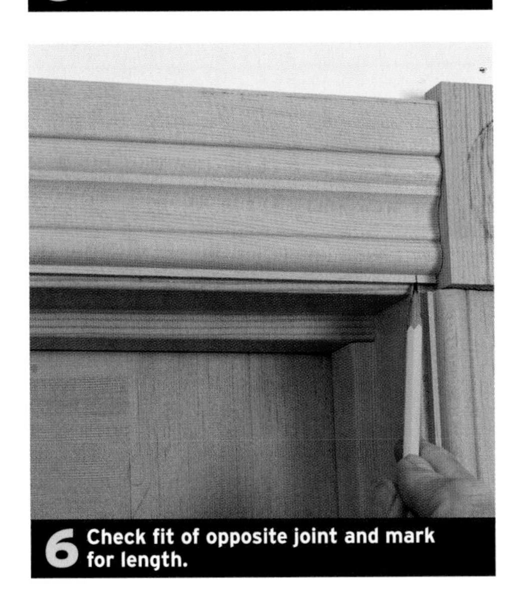

6 Check fit of opposite joint and mark for length.

RABBETED BANDING

1 Use miter joint on casing to mark banding to length.

Rabbeted edge banding, like other rabbeted moldings, is best marked to length in place. When used over mitered casings, the easiest place to mark the banding to length is on the inside edge of the banding where it intersects the miter joint of the casing below it ❶. When attempting to tighten the banding to the face of the trim nail through the face of the banding ❷. If the banding lays tight to the casing, nail it into the side of the casing where the nails will not show as readily ❸.

2 Face nailing tightens gap between banding and casing.

3 Side nailing makes nails harder to see.

RUNNING MOLDINGS

RUNNING MOLDINGS ARE CONTINUous, horizontal moldings and include crown molding, picture rails, chair rails, bandboards, baseboards, and wall caps. The installation techniques for running moldings are similar, with the exception of crown molding. Basically, running moldings fall into one of three different categories: The first, "Wall Surface Moldings," include all moldings installed flat on the wall surface such as baseboard, band boards, and chair and picture moldings. The second category, "Angled Transition Moldings," are moldings installed at an angle to the wall surface and include crown and some bed moldings. Finally, "Flat Laid Moldings" are installed primarily as caps on half walls.

Crown and baseboard molding serve the practical purpose of concealing the joint between the wall and the surface that intersects with it. Chair rail was once important for protecting a wall from chair backs rubbing against it. Nowadays, it's usually a cap for wainscot. Other types of running moldings are more purely decorative, but in every instance they serve to tie a room together and define its style.

BASEBOARD MOLDING

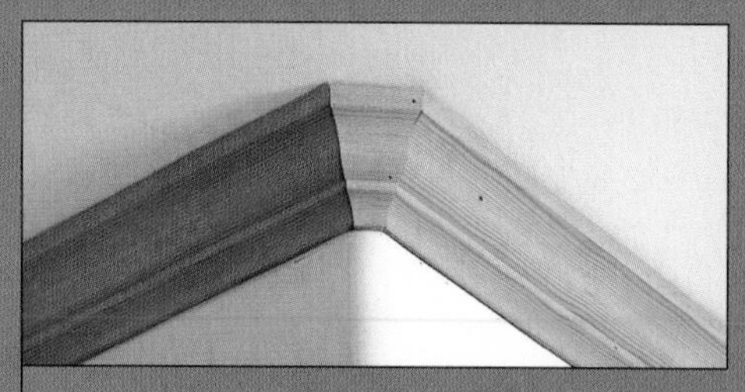

CROWN MOLDING

WALL CAP

PERIMETER BAND BOARDS

Photo by Randy O'Rourke

design options

BASEBOARD

A multitude of possibilities

As its name implies, baseboard is installed at the base of a wall. Its practical purpose is to prevent damage to the bottom of the wall from feet and furniture. Aesthetically, baseboard covers the seam and provides a visually satisfying transition from the horizontal floor to the vertical wall. On a hard surface floor the base is installed tight to the flooring while on a carpeted floor the base is held off the subfloor (usually ½ in.) to allow the carpet to be tucked under the base.

Multiple piece baseboards are a combination of moldings combined to create a larger more elaborate assembly and are commonly found in more traditionally styled homes. In addition to creating a more interesting looking baseboard the additional moldings can hide uneven wall and floor surfaces. A common three piece baseboard design consists of the baseboard itself (which often is simply a piece of dimensional stock), a small molding cap that sits on top of the base, and a shoe molding that covers the juncture of the baseboard and the floor. These smaller secondary moldings serve to cover up gaps between the wall and floor and the baseboard.

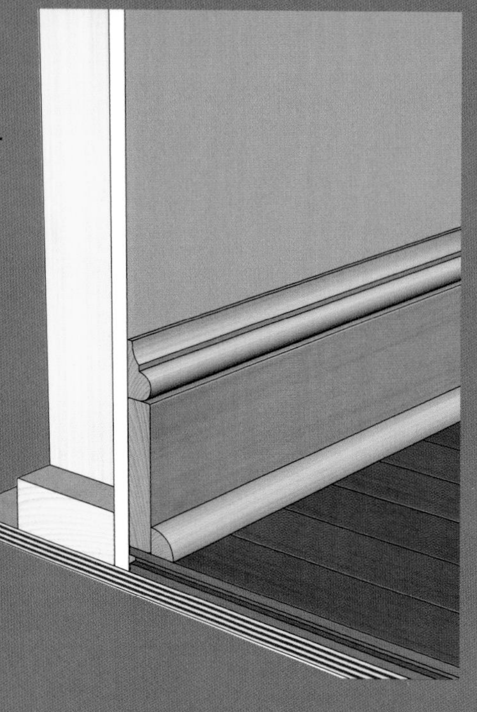

ADD A BEAD

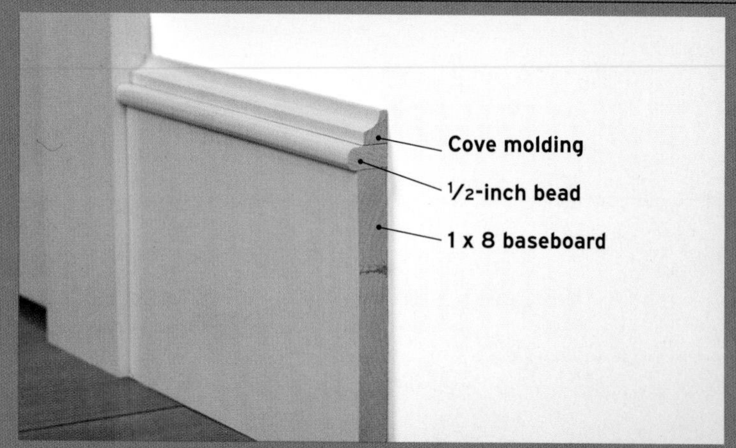

Add a shopmade bead under the cove molding for a more formal look.

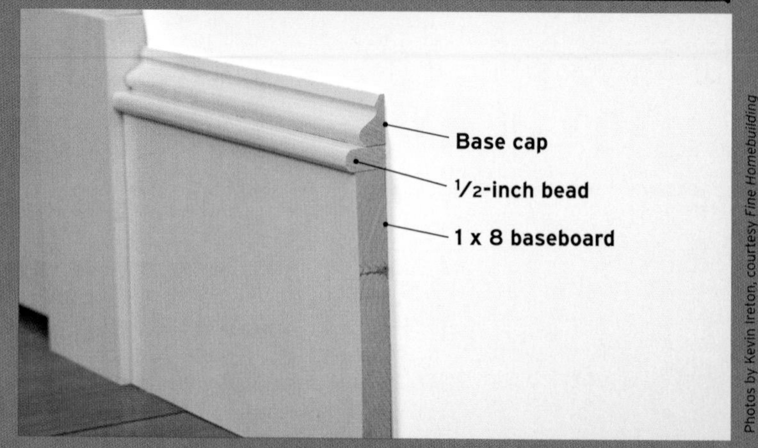

Adding a bead between the cap and the base gives formality to baseboard.

Photos by Kevin Ireton, courtesy *Fine Homebuilding*

TRY DIFFERENT BASE CAPS

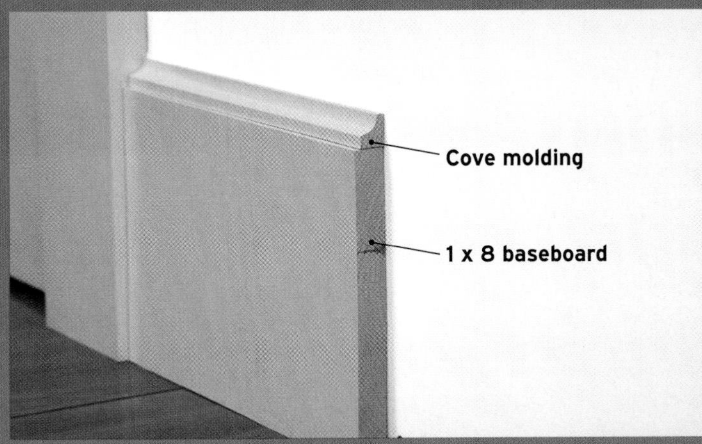

A basic 1 in. x 8 in. baseboard with stock cove molding. You could also use 1 in. x 4 in. or 1 in. x 6 in. in the same way.

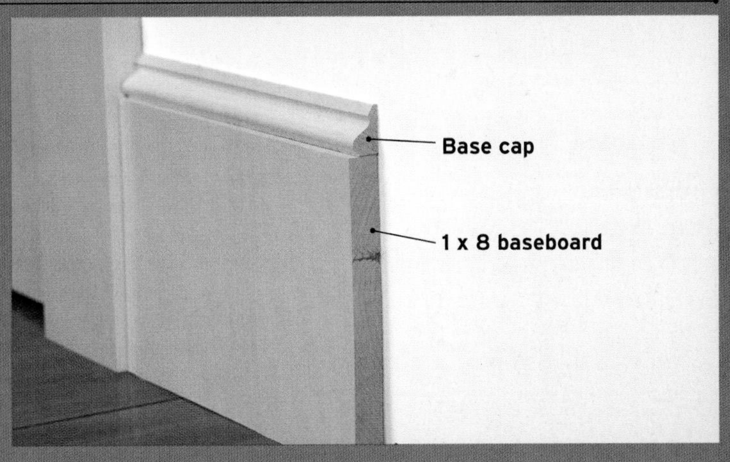

Use a simple base cap for a clean classical design.

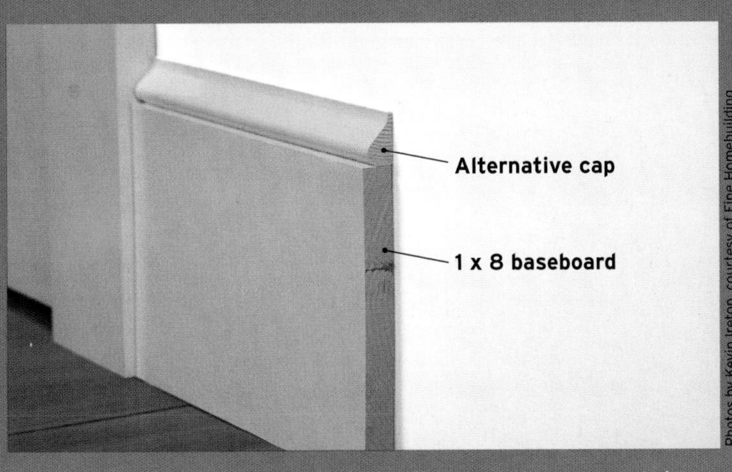

This stock molding is more rounded than the simple base cap and gives a softer appearance.

Photos by Kevin Ireton, courtesy of Fine Homebuilding

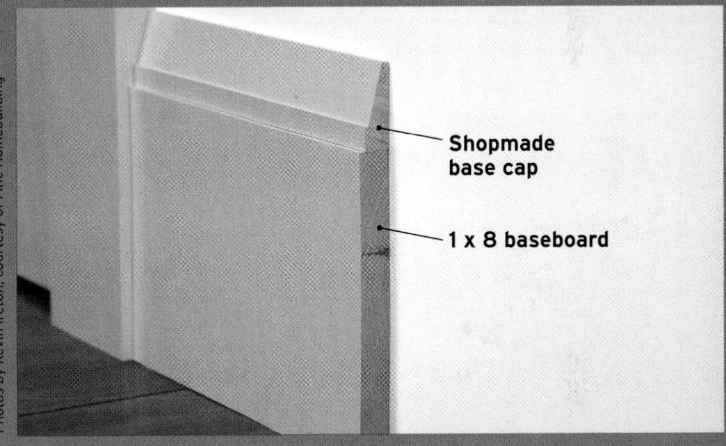

This shopmade base cap was cut on the tablesaw. It works with both colonial and Craftsman styles.

FURR IT OUT FOR A MORE FORMAL LOOK

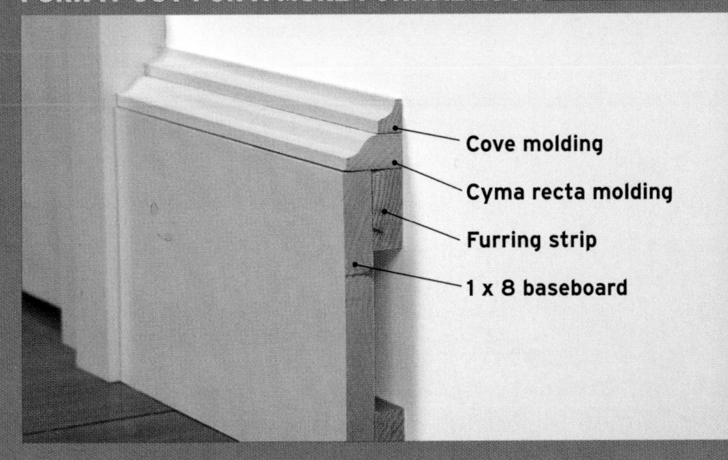

Furring out the baseboard from the wall creates space for a cyma recta molding under the cove cap.

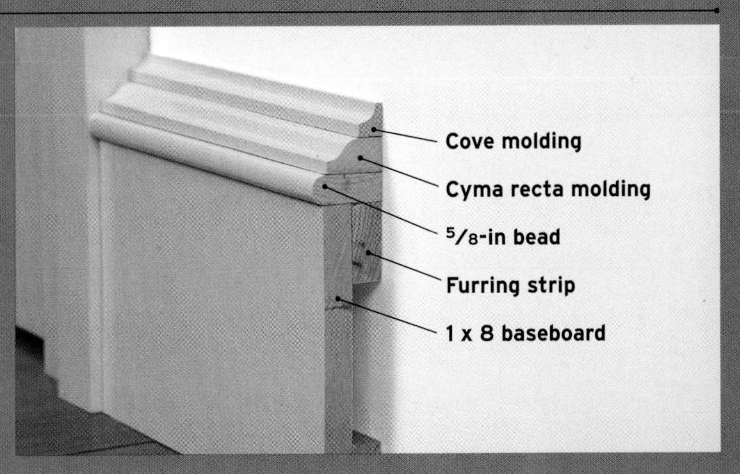

For a high style look, add a 5/8-in.-thick bead under the cyma recta molding and cove cap.

PLANNING THE INSTALLATION

Good planning will ensure both the best use of materials and save you time. It's not necessary to work your way around the room in any particular direction but rather concentrate on getting as many pieces installed or measured per trip as you can.

Minimize situations where you have to create finished joints at both ends of a molding. This may be unavoidable when you have to install a molding between two already installed moldings or when you have to run between an installed molding and an outside corner.

Always strive to combine measuring and cutting as many running-molding pieces as possible per trip. On the initial trip to a room measure the rough lengths for the outside corners and all the pieces that will butt into standing moldings. If there are long walls that will require spliced moldings, locate a stud and take the measurement of the first piece needed. Cut the adjoining piece to rough length (you'll mark it in place for exact length).

Use two matching 30-degree angles for splicing wall moldings. With the exception of narrow baseboard, which can be spliced and nailed anywhere by using the continuous bottom wall plate to nail to, all the baseboard splices are centered over a wall stud with the ends of both pieces being nailed to it. After cutting and nailing these in, measure and record the lengths of all the pieces butting into them and at what end the joint is to be cut. On pieces that join in finished trim, cut the pieces 1⁄16 in. or 1⁄8 in. longer and spring them in place. The exception is trim butted to a window or door casing. Springing can move moldings out of place and can affect the reveal. Cut the piece to the exact length.

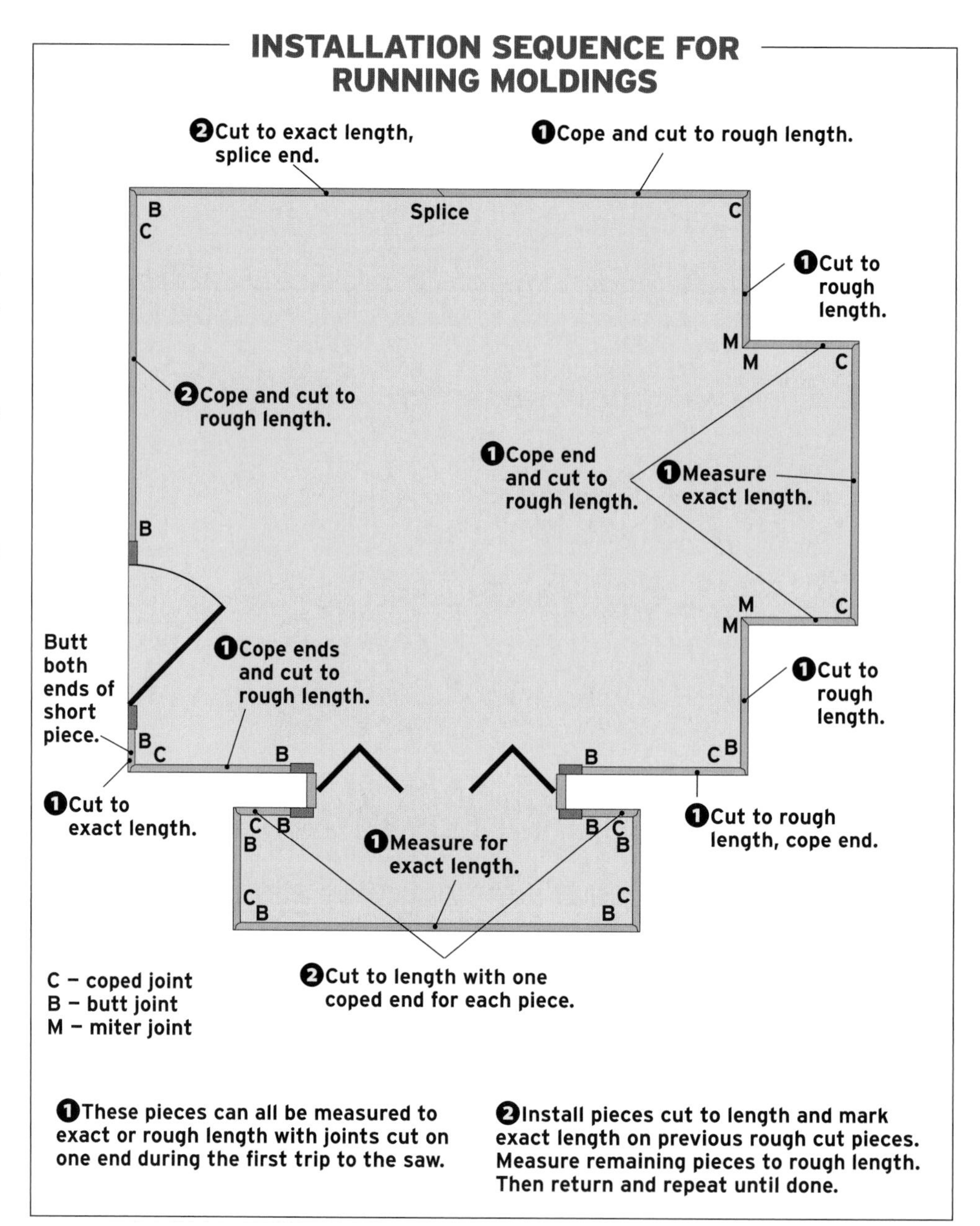

Rules of Thumb for Running Moldings

- Minimize situations where you have to create finished joints at both ends.
- Mark directly on the pieces to finished length.
- For spliced moldings, locate a stud before cutting pieces to length.
- Use two matching thirty-degree angles for splicing wall moldings.
- Except for trim butted to window or door casing, spring sections in place.

INSTALLING BASEBOARD AROUND 90-DEGREE CORNERS

Using the wall's outside corner as an index, hold a length of baseboard against the wall and mark the inside corner of the miter joint with either a pencil or, for more accuracy, a utility knife ❶. Use the empirical method of trial and error to determine the angle of the miter cuts—with a little practice you'll achieve the correct miter angle the first time: Cut two test pieces of molding at 45 degrees and check the fit of the corner miter. If the joint is not correct, adjust the angle cut on the miter saw accordingly and recut the miters until they fit. Once you've achieved a tight fit, apply yellow glue to the mitered surfaces and nail the baseboards to the wall ❷. Secure the miter joint itself by pin-nailing across the corner into both halves of the miter joint ❸.

1 Mark the length of the base with a utility knife.

2 Apply glue to the miters and nail the base to the wall.

3 Use a pinner to cross-nail the miter joint.

Flattening Bowed Baseboards **An effective way to straighten a bowed baseboard is to use your body weight while nailing it in place. Set one end of a piece of board about 3 ft. long on top of the base (exercise some caution here as it can slip off the top with you on it). Set the other end on the floor. Now kneel on the board using your weight to bend the base down and nail it off once it is pushed tight to the floor.**

INSTALLING BASEBOARD AROUND RADIUSED CORNERS

Radiused outside corners require a transition piece to carry the molding around the corner. In the case of baseboard a pre-made radiused corner block can be used **A**. These blocks are available in most of the standard base profiles and are nailed to the corner first with the base pieces glued and butt jointed to them. Inside corner joints can be butted to one another or a corner block, mitered or coped **B**.

See "Installing Corner Blocks," p. 122.

You can also make up a transition piece from the base material to install baseboard around a radiused corner. First, build a corner jig, to determine the length of the abutting base pieces to the transition piece, because you'll install them before the transition piece. Hold the jig in place and then draw a line on the wall at the end of each leg **1**. Determine the length of the adjacent pieces of base by adding the length of the jig leg to the length of the base as measured to the jig or the mark. Cut a 22½ degree miter on the end of the first base piece and nail it to the wall. Note that on standard ½ in. drywall corners the transition piece is ⅝ in. long from short point to short point **2**. Now apply construction adhesive to the wall corner **3** and yellow glue to the miter joint and pin the corner piece to the previously installed piece of base **4**. Apply glue to the next miter joint and nail the second base leg to the wall framing. Finish by pinning the miters together **5**.

A A factory made transition piece bridges wall moldings around a radiused corner.

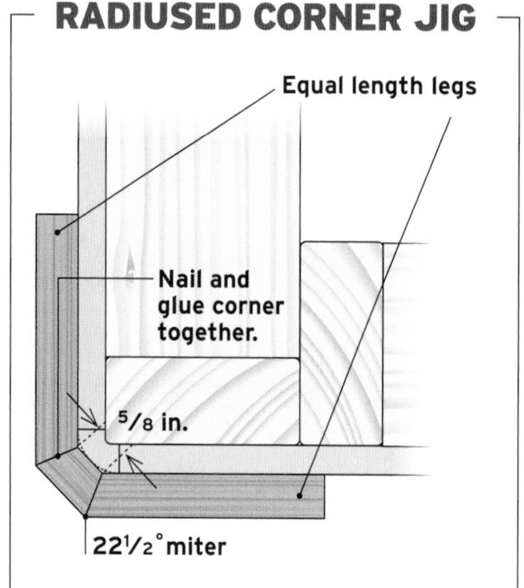

1 Mark the wall using a corner jig and measure to mark at end of each leg.

3 Apply a bead of construction adhesive to the corner.

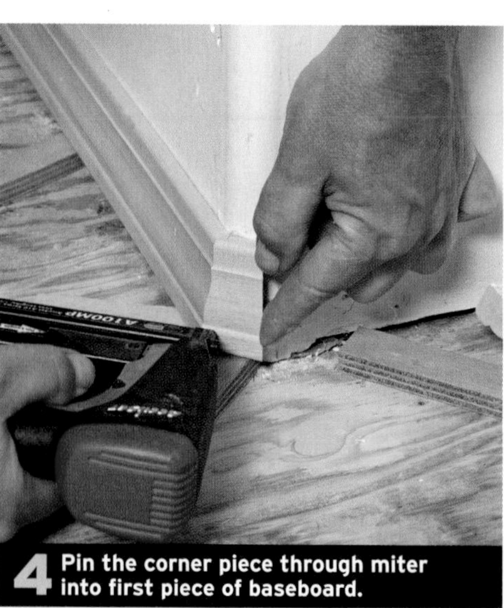

4 Pin the corner piece through miter into first piece of baseboard.

B A traditional styled corner block can be installed on bullnosed corner.

2 Corner transition piece is 5/8 in. short point to short point dimension.

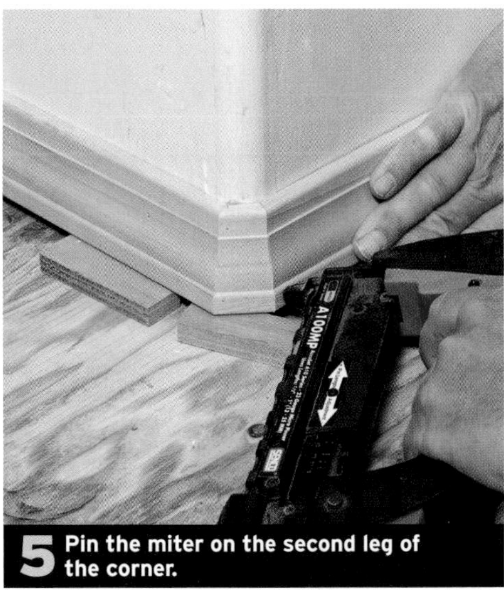

5 Pin the miter on the second leg of the corner.

COPING INSIDE CORNERS

Use an inverted saber saw or, alternately, a coping saw to cut a cope joint on baseboard. The saber saw is faster, though a coping saw will get the job done. If your saw has an orbital function, it may help to shut it off to make the cutting action less aggressive and more controllable. Start by marking the length measurement on the piece and then make a forty-five degree miter cut, locating the short point of the miter on the line. Now cut out the waste from the molded profile ❶, slightly undercutting the line will ensure a tight fit. If you are not comfortable using a saber saw or one is not available, use a coping saw.

On larger or more complicated profiles, it may require cutting in from both directions and making multiple passes in order to remove all the waste ❷. A bit of chiseling or filing may also be required if the profile is extraordinarily intricate.

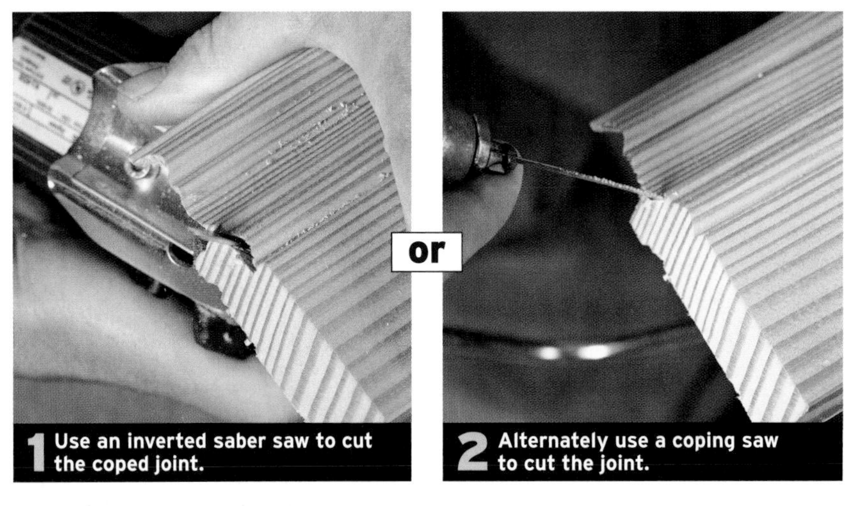

1 Use an inverted saber saw to cut the coped joint.

2 Alternately use a coping saw to cut the joint.

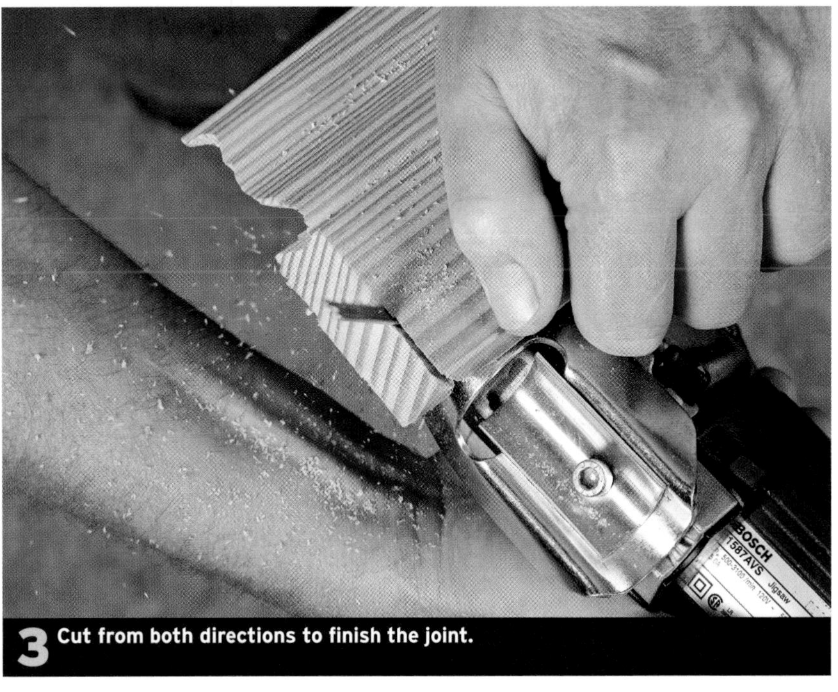

3 Cut from both directions to finish the joint.

INSTALLING CORNER BLOCKS

1 Apply construction adhesive to back of corner block before installing.

2 Glue the ends of the abutting baseboard before installation.

3 Butt the base pieces to the corner block and nail in place.

4 Installing an outside corner block to the wall corner.

Corner blocks add a decorative element to inside baseboard corners and eliminate the need to make challenging and time-consuming cope joints. At an outside corner, a block eliminates the need for a miter joint. When installing an inside corner, you will find it easier to get a tight fit to the corner block if it is not nailed securely to the corner. Instead, apply a bit of construction adhesive to the back two faces of the corner block and press the block into the corner ❶. The construction adhesive allows a little adjustment of the corner block if the joint needs to be tightened up. Cut the abutting pieces of baseboard to length and apply yellow glue to the ends of the butt joints ❷ and nail the pieces in place ❸. Install an outside corner block to the wall corner and then cut the abutting baseboard to length, gluing the butt joints. ❹

SPLICING BASEBOARDS

Splices on baseboards smaller than 2½ in. in height do not need to be located over a wall stud as they can be nailed adequately into the bottom wall plate. On anything higher, however, you should center the splice joint on a wall stud. Start by locating and marking on the wall the location of the stud where the joint will be. Measure the length of the first piece and cut a 30-degree miter with the face of the miter showing on the end of the piece. Center the miter on the stud location and nail it to the wall ❶. Measure the length of the second piece and cut a matching 30-degree angle on the adjoining end. Apply glue to the miters before installing ❷. Nail the second piece of base to the wall through the miter, placing one nail into the plate and another into the wall stud ❸.

1 Nail first piece with 30-degree angle cut on end into stud.

TIPPED BASEBOARD

Tipped baseboard occurs if the drywall stops short of the floor. To correct it, drive a drywall screw into the wall behind the base to straighten it out before applying the intersecting piece.

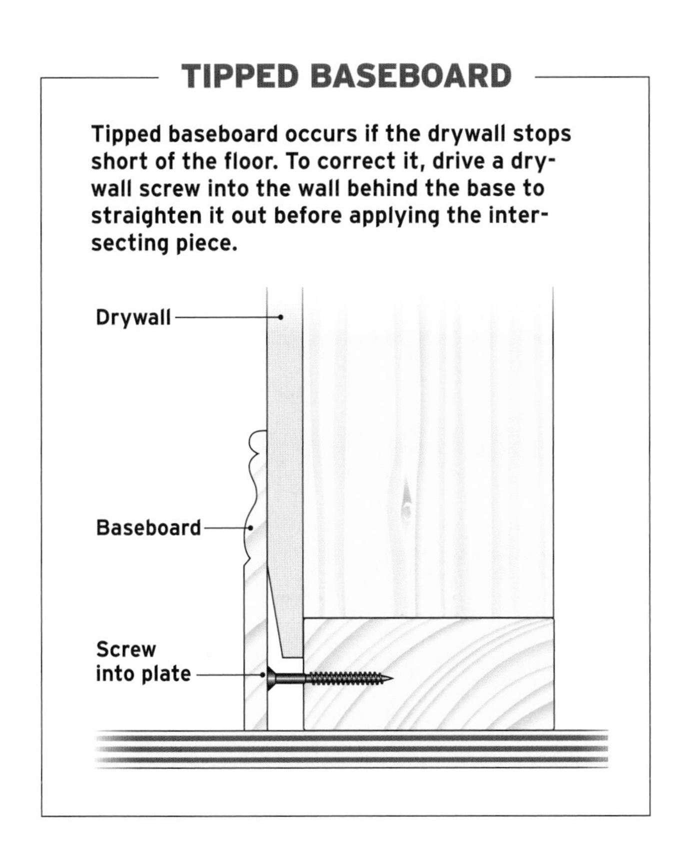

2 Apply glue to miter.

3 Nail second piece into both the wall plate and the stud above.

SCRIBING BASEBOARD TO THE FLOOR

1 Use shims to level baseboard if floor is not level.

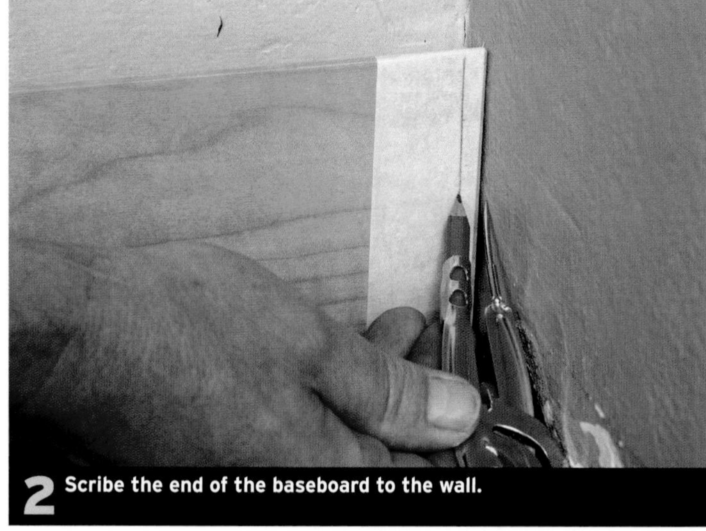

2 Scribe the end of the baseboard to the wall.

3 Mark the base to length using a preacher against standing molding. .

4 Scribe floor profile onto base with compass.

5 Cut to the scribed line with an inverted saber saw.

6 Trim baseboard to the line with a belt sander.

or

Alternatively use a block plane for trimming to line.

If the floor is uneven or out-of-level, you will need to scribe it to the floor so that the top edge of the baseboard will be level. Begin by creating level shim points for the base to set on during scribing ❶. Cut the base a little longer than required and scribe the first end to the abutting wall, if necessary ❷. Use a preacher to mark the cut line for the other end where it abuts standing molding ❸. The piece should now fit tight at its ends while sitting level on the shims. Using a scriber, transfer the outline of the floor onto the base ❹. Cut with an inverted saber saw being careful to leave the line ❺. Create an under-bevel by slightly tilting the saw when cutting or while finish-trimming with a belt sander ❻ or with a block plane if the line isn't too irregular.

To prevent the board from jumping around while working on it, place the board across two sawhorses or clamp it to a bench top being careful that the cut line overhangs far enough past the edge of the bench to allow room for the saber saw.

INSTALLING MOLDINGS ON UNEVEN WALLS

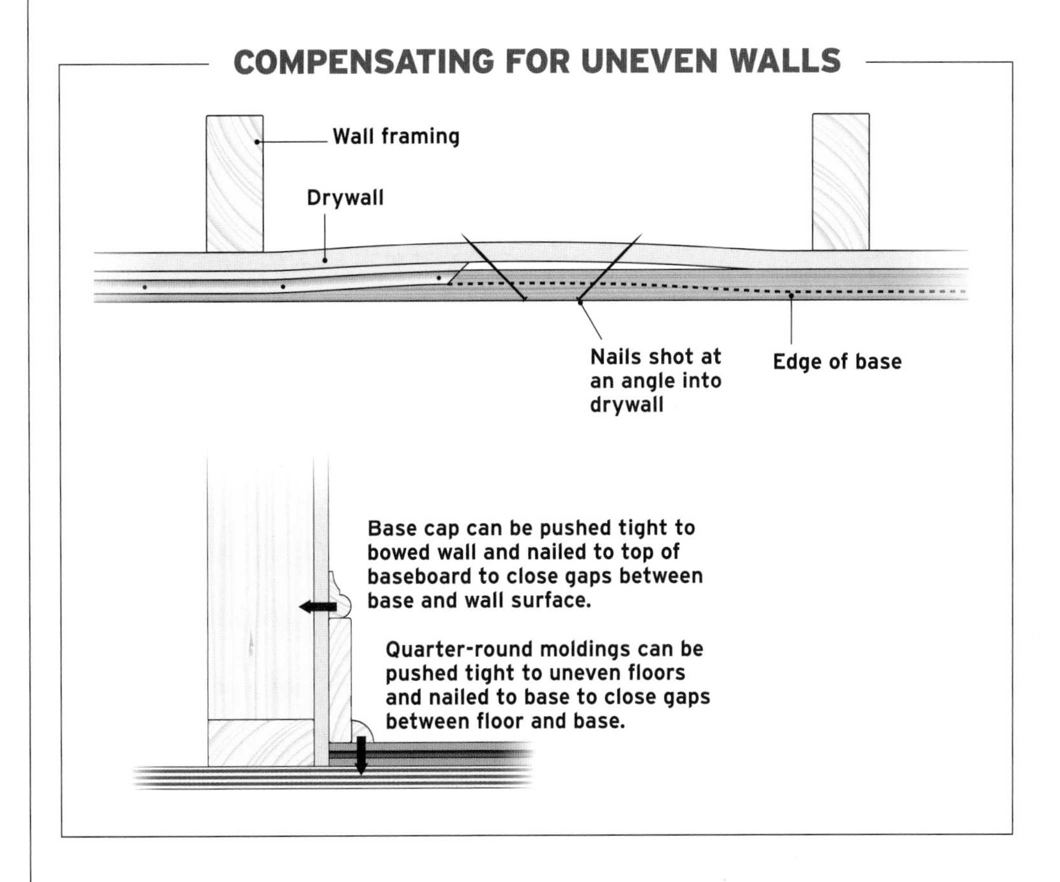

Installing flat wall moldings to a wall that has dips or humps can be a challenge. Paint grade moldings are the easiest to install because the gaps can be caulked and painted. Due to their greater flexibility MDF and urethane moldings are more easily bent to an irregular wall than solid wood moldings. Several nails shot through the base into the Sheetrock at an angle combined with a bead of construction adhesive applied to the wall will keep them tight to the wall in between the studs. Caulking works fine for filling in gaps but will not keep the base tight to the wall. The same is true for natural wood moldings in the smaller dimensions (less than 5/8 in. thick), but greater care is needed when caulking the top. In the case of multipiece baseboards, a base cap can be pushed tight to the wall helping to close the gap before being nailed tight to the top of the baseboard.

TRADE SECRET

When marking the scribe line onto prefinished, dark or coarsely textured woods, first apply a piece of masking tape to the board to make the line clearly visible and to help prevent the pencil from being diverted by deep grain lines.

TRANSITIONING MOLDINGS

When a running molding intersects itself, the key to good transition is usually an accurately cut joint. In the case where a molding transitions to another type of molding or changes levels, it may take some finessing to make the result look pleasing and finished. The situation where a molding intersects with window or door frames can also present some challenges.

Transitioning to standing moldings

Often chair or picture rail feature a bed molding or transition molding installed directly under it. When running moldings intersect standing moldings, they will often stick out past the surface of the standing molding and either have to be notched out to fit over the molding or beveled at the end tapering into the molding.

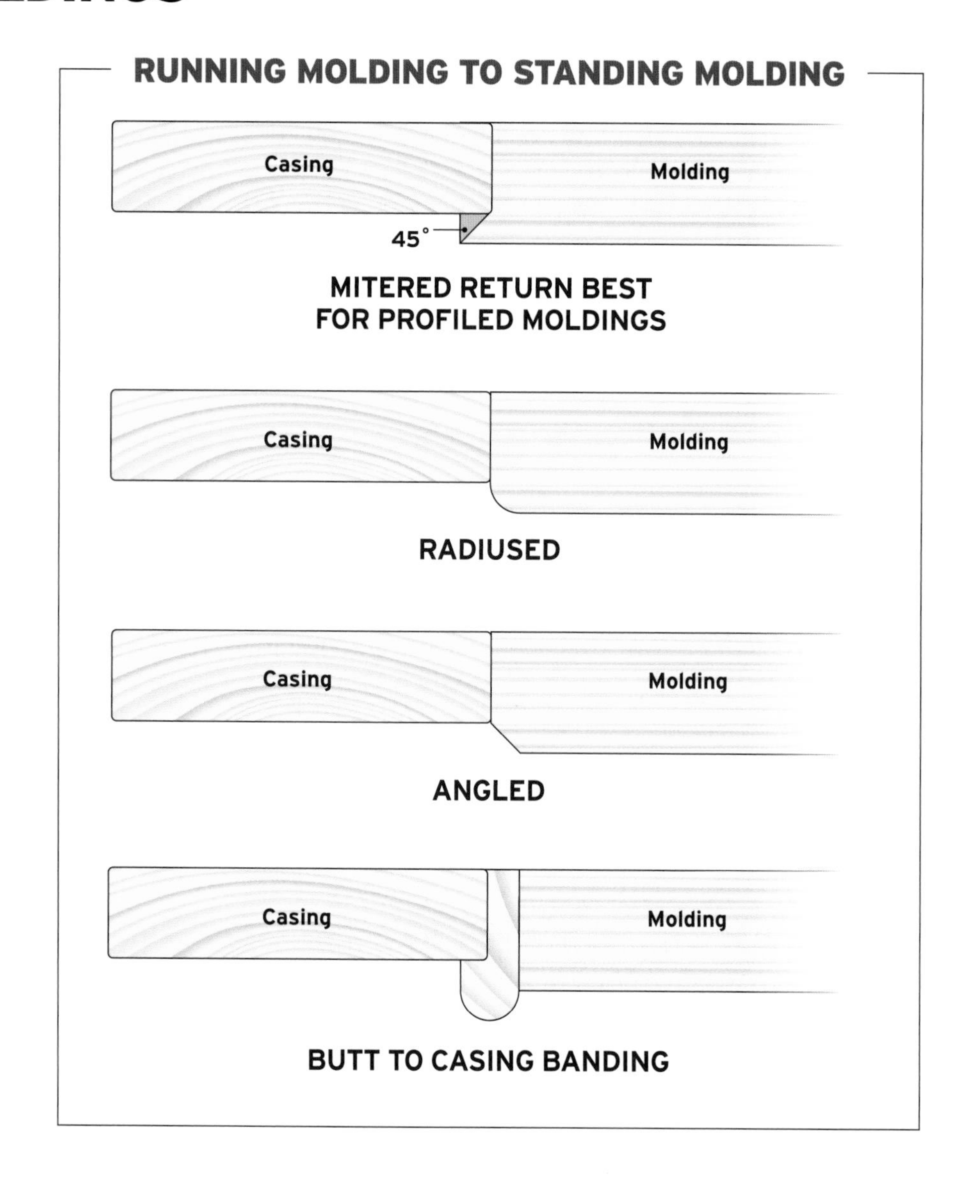

RUNNING BASEBOARD TO STAIR SKIRTBOARDS

There are many different methods of joining baseboards to skirtboards. The hard part is getting the large flat surface of the skirtboards to blend gracefully into the base. When using flat dimensional baseboard of the same thickness as the skirt material the skirtboards can flow right into it and produce a clean look. If either material is different thickness or the base is molded, there will need to be a transition. The edges of the thicker piece can simply be eased over with the thinner one butting into it. This works when butting a thinner base or a molded base into the skirt.

Another option is to use transition blocks between the two pieces. These are slightly larger than either piece and tend to draw attention to the joint rather than blend in. A problem with the base/skirt intersection appears when the long point of the skirt does not end up at the same height as the top of the base. At the bottom of a stair, the skirt can be clipped off at whatever height is necessary to match the height of the base. At the top of a stairs, this can usually be corrected by either clipping off the top of the skirt to match the base if the base is too low or if the base is too high notching the skirtboards past the edge of the top riser allowing it to continue until it reaches the correct height.

BASEBOARD-SKIRTBOARD VARIATIONS

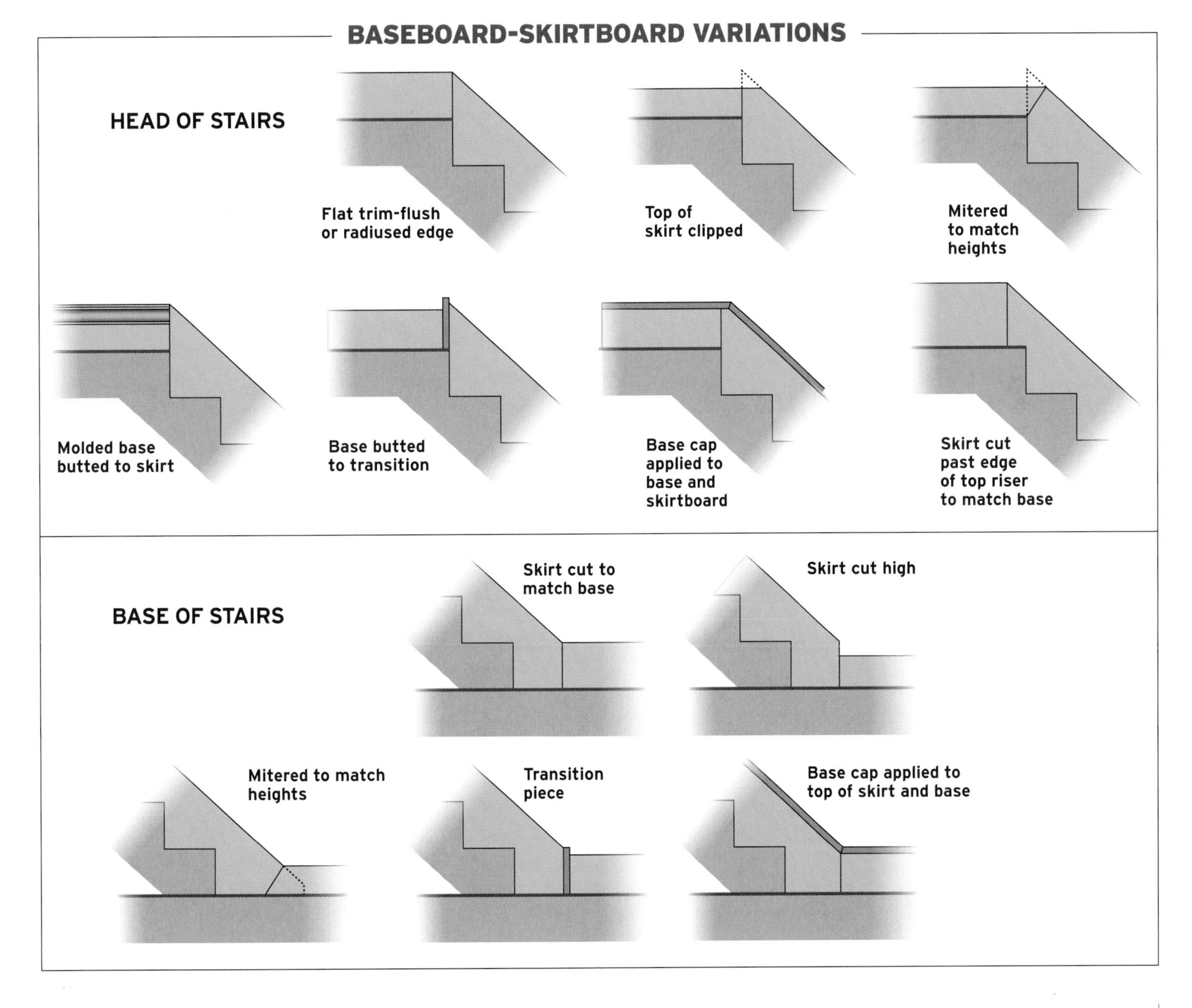

CROWN MOLDING ON A 90-DEGREE CORNER

Outside corner joints are the most troublesome in crown installation. Make a set of test pieces and check the fit for each inside and outside corner before cutting the final pieces to length. The fit of the miters can be adjusted by slightly moving the material at the joint up or down or by shimming behind the miters to close a gap. Once the pieces fit, the miters are glued and cross nailed. Don't nail close to the corner until both pieces are at least tacked up as the joint may need to be manipulated by shifting the moldings slightly up or down.

Start installation by chalking a line on the wall to indicate the bottom edge of the crown. An alternative to measuring down from the ceiling to obtain the bottom edge of the crown molding is to use a laser level to mark level points at a convenient height around the room perimeter. Then measure from these points up to the desired height for the bottom of the crown molding. Now start outside corner installation by determining the corner angle. Cut 45-degree miters on opposite ends of two pieces of scrap. Make the pieces long enough to enable the crown to get past any swelling of the corner due to drywall compound build up and provide an accurate angle reading. Align the test pieces to the positioning line and check the fit of the miters ❶. Adjust the angle cuts as necessary until the joint fits. Now hold the individual pieces in position and mark them to length. Do them both now as once the first piece is up the second can no longer be marked in place ❷. Cut a miter at the mark and nail the first piece in position. Keep nails several feet away from the corner so you can adjust the fit of the miter if necessary ❸. Now check the fit of the test piece to the installed piece ❹. Correct the miter if necessary and cut this angle on the finished piece once again checking the joint before installing ❺. Relieve the backside of the miter joint with a block plane if necessary to close the miter ❻. Now nail the piece of crown on the line ❼ and glue and cross nail the miters ❽. Nail into the top wall plates and whenever possible the ceiling joists. Long screws can also be driven through the crown into the top plates and the holes either being puttied or plugged depending on finish.

On paint grade moldings nails can be shot through the top of the crown into the ceiling Sheetrock at opposing angles with adhesive caulking applied along the joint.

Blocking for large crown

Unless blocking was preinstalled, when installing crown parallel to the ceiling framing there will not be any place to nail the top of the crown into the ceiling. If the crown is larger than 4 in. you may have to install angled blocking behind the molding to provide for nailing.

WARNING Installing long lengths of crown molding requires standing on ladders and is more easily and safely done with the help of another person—even if all they do is hold up the other end of the board. If no one is available, you can use extendable floor clamps to hold the crown in place or you can support the other end of the board by resting it on a nail driven into the Sheetrock.

CORNER BLOCKING FOR CROWN INSTALLATION

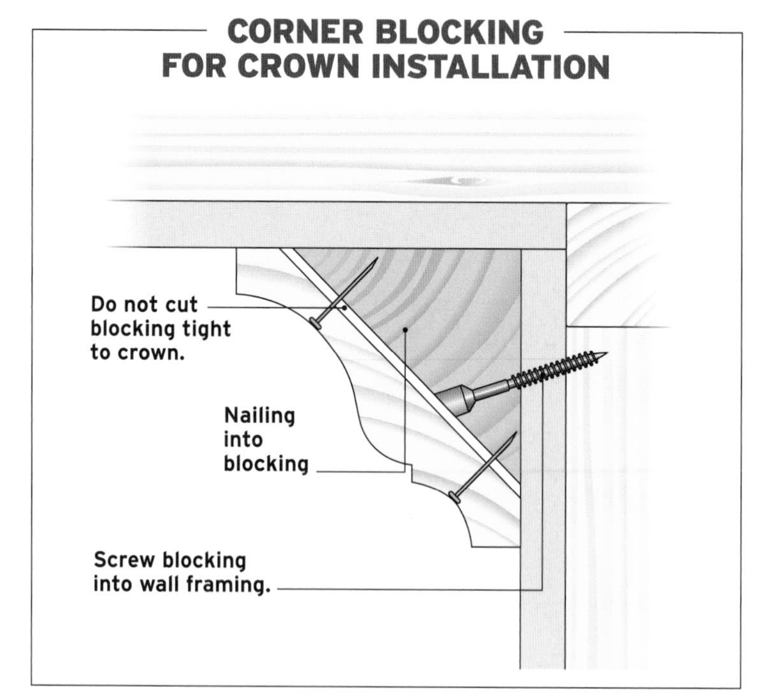

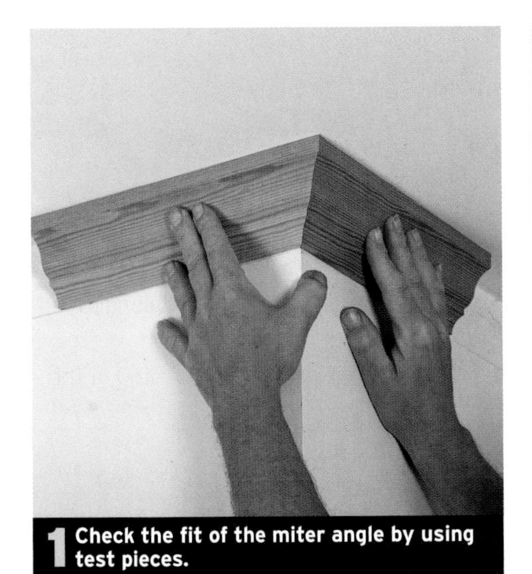

1 Check the fit of the miter angle by using test pieces.

2 Mark the pieces to length.

3 Nail the first piece to the wall.

4 Use the test piece to check the fit of the joint.

5 Check the fit of the miter before installing the second piece.

6 Relieve the backside of miter with a block plane if necessary.

7 Nail the second corner piece to wall.

8 Nail the mitered corner together.

or

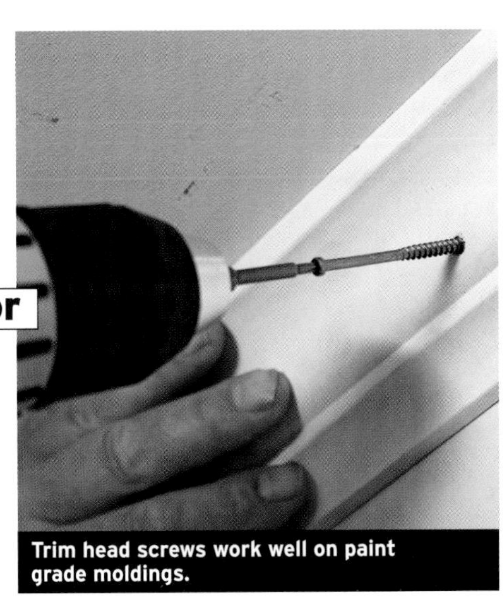

Trim head screws work well on paint grade moldings.

CROWN MOLDING ON A RADIUSED CORNER

Crown moldings are installed around bullnosed corners in much the same way that baseboard is. As these corners always need a bit of adjusting even if you know the exact corner angle plan to use the empirical method of trial and error to fine-tune the corner miters. After marking the bottom edge location lines on the wall assemble a test corner with twenty-two and one half degrees on all the miter cuts. Note that a short point dimension of 5/8 in. for the transition piece works on commercially available radiused wall corner bead ❶. Make the legs of the test piece equal length so you'll only have to work with one measurement to add to the length of the converging moldings. You'll use this pre-assembled corner to check the fit of the miter angle and to mark the wall at the end of each leg to determine the length of the sidepieces ❷. Begin by holding the test corner in place and noting which way the miter cuts need to be tweaked to make the corner fit ❸. Adjust the miter setting on your saw accordingly and then try the new miter angle on some pieces of scrap before cutting the final pieces. Once you've determined the correct miter angle, find the length of the long wall pieces by measuring to the previously marked line indicating the end of the test piece (or measure to the end of the test piece itself) and add that measurement to the leg length of the test piece ❹. Align the first piece of crown to the location line on the wall and nail in place ❺. Nail the second length of crown in place leaving the last several feet before the corner loose in order to use it and the transition piece to fine-tune the corner assembly. Then glue and nail the transition piece through the miters into the side pieces ❻.

1 Assemble a test corner.

2 Use the test corner to determine if the corner angle needs adjustment

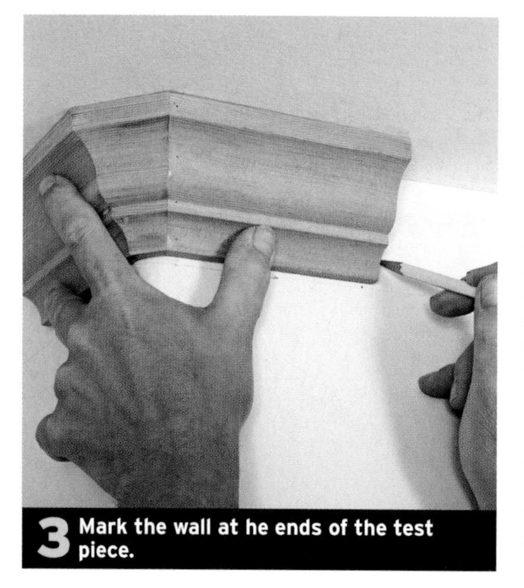

3 Mark the wall at he ends of the test piece.

4 Measure to line marked on wall or to test piece itself.

5 Nail the first piece of crown to wall.

6 Once sides are installed nail transition piece through miters to sides.

COPING INSIDE CORNERS

Inside corners are either mitered or coped. I will always cope when given the choice as it produces a tighter, longer lasting joint. With paint grade material mitering can work as the joints may be caulked but on stain grade coping is the best way to it. Start by using a chop-saw to cut a 45-degree angle on one end of the crown molding ❶. This exposes the edge profile and provides a line to cut to. Remove the waste with an inverted saber saw (for best results, fit the saw with a Collins coping foot) ❷.

As an alternative—or if the back cut required is larger than the capacity of the saber saw—use a coping saw. Since the coping saw is making a rip cut (rather than a crosscut) as the back cut angle increases, it will require a bit more time and effort. Both tools will require that the cut be made in multiple passes starting from both directions. If the joint was cut accurately but still will not fit tight, remove more stock from the back of the cut.

1 Crown molding cut in proper position prior to coping.

2 Saber saw is partway through a coping cut.

or

As an alternative use a coping saw to cut cope.

CUTTING CROWN IN POSITION ON A MITER SAW

To cut crown on a miter saw you'll place it upside down in its spring angle position on the saw table. Since most miter saws do not have a fence high enough to hold crown molding at their spring angle, you'll need to add a piece of wood to the factory fence. Measure the maximum height fence your saw can handle and then rip a board to that height. Attach it to the existing fence with short screws from the back ❶. Now use a piece of crown molding and a framing square to determine how high up the fence to locate the indexing line that the crown will be held to for cutting ❷. Measure this distance from the base of the saw and use a straightedge to draw a line on the auxiliary fence ❸. The line provides a positive visual location to align the molding to while cutting the various joints ❹, ❺.

1 Attach an auxiliary fence to factory fence.

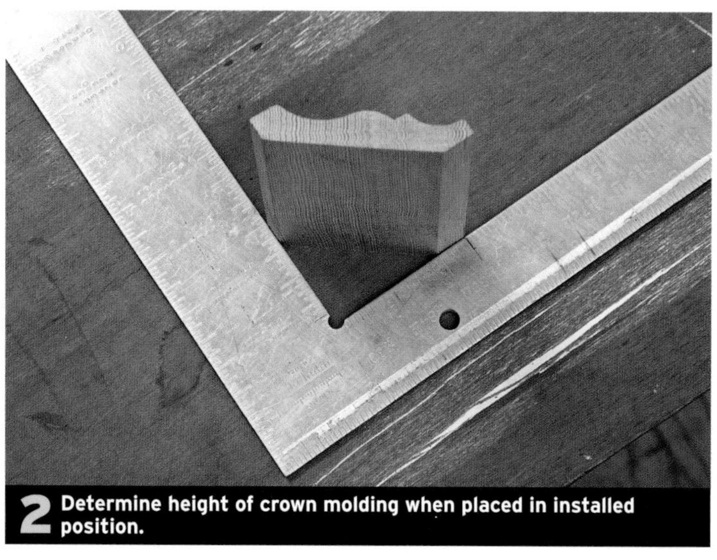

2 Determine height of crown molding when placed in installed position.

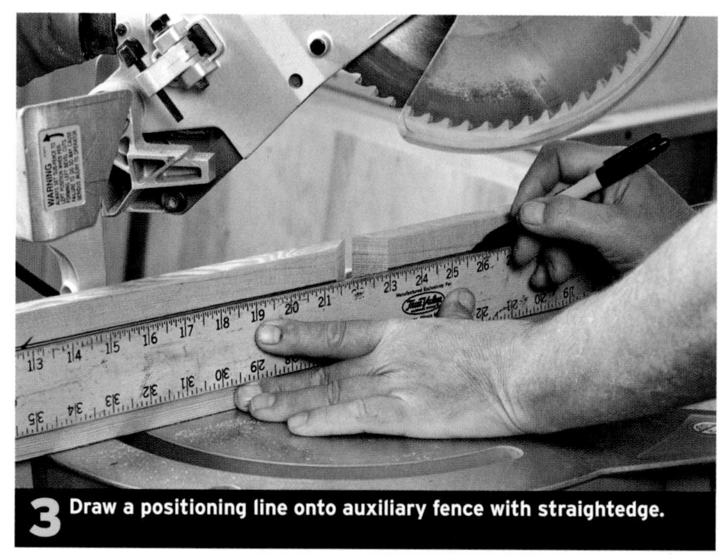

3 Draw a positioning line onto auxiliary fence with straightedge.

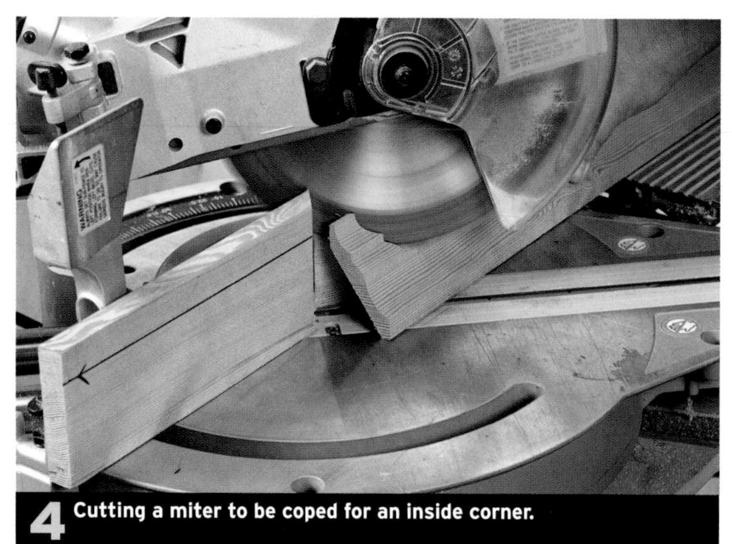

4 Cutting a miter to be coped for an inside corner.

5 Cutting a miter for an outside corner.

DETERMINING AND TRANSFERRING ANGLES

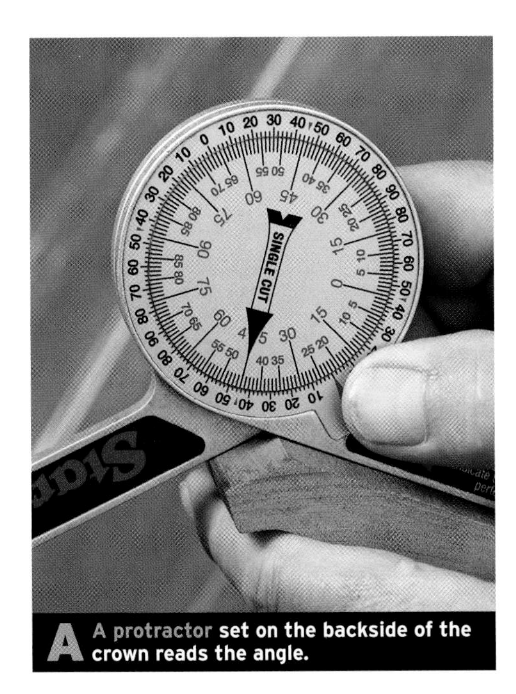

A A protractor set on the backside of the crown reads the angle.

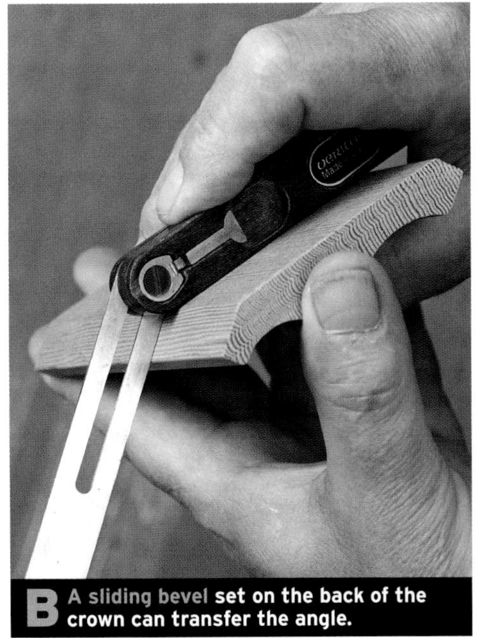

B A sliding bevel set on the back of the crown can transfer the angle.

Unlike a standard chopsaw, a compound miter saw can cut crown molding lying flat. Because the crown is not positioned at its spring angle, however, it's a little harder to keep the orientation of the cuts straight in your head—at least at first. You need to work from a chart of angle settings to set the cuts and you must know the spring angle of the crown you are installing. To determine the spring angle, hold a bevel gauge tight to the back and the bottom edge of the crown Ⓐ.

Alternately, the angle can be copied with an adjustable square and read off a bevel board or protractor Ⓑ. Once the spring angle is known go to the chart below, and set the miter saw accordingly.

SPRING ANGLE

Crown moldings and some bed moldings are designed to sit at an angle or "spring" out from the wall and intersect the ceiling. The angle that the crown comes off the wall is called the spring angle and is listed on cutting charts to set up blade angles for cutting crown moldings flat on the saw table. The most common spring angles are 38 degrees and 45 degrees.

ANGLE SETTINGS FOR CUTTING CROWN FLAT

Type of Crown (spring angle in degrees)	Miter (angle on table in degrees)	Bevel (tilt of blade in degrees)
30	27	38
35	30.5	35
38	31.5	34
40	33	33
45	35	30
52	38	26

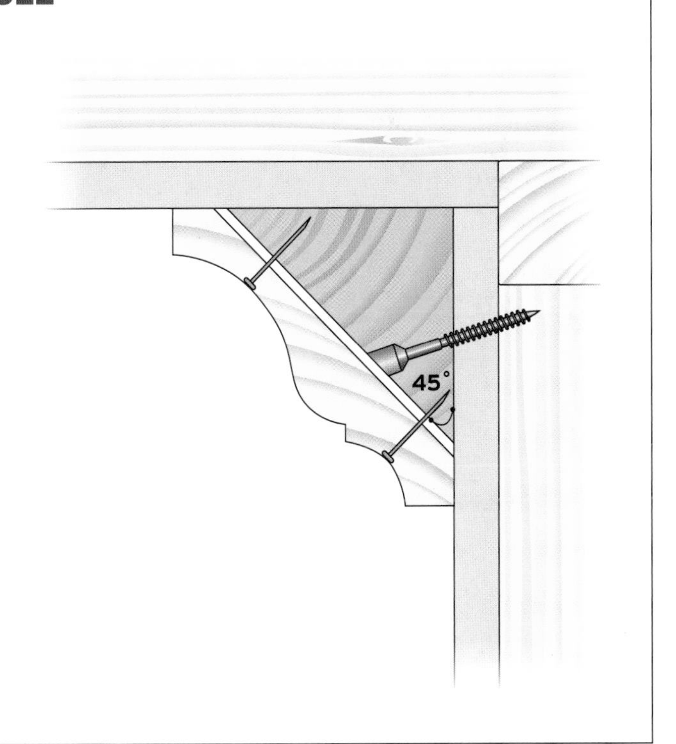

CUTTING CROWN FLAT ON A COMPOUND MITER SAW

OUTSIDE CORNERS

Left-hand mitered outside corner

For the left-hand piece, place the piece to the right of the blade, with the bottom edge against the fence. Miter angle is set to the left of center.

Left-hand mitered corner in position.

Right-hand mitered outside corner

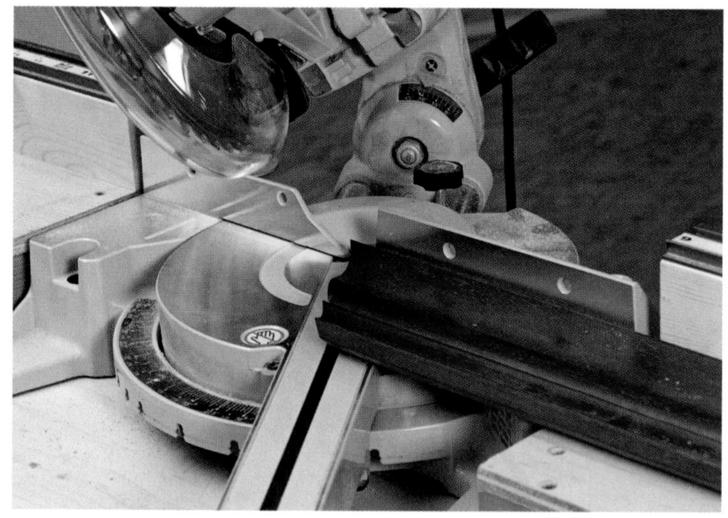

For the right-hand piece, place the molding to the right of the blade, with the top edge against the fence. Miter angle is set to the right of center.

Right-hand mitered corner in position.

INSIDE CORNERS

Left-hand inside corner

For a left-hand coped piece, place the piece on the left of the blade with the top edge against the fence. Miter angle is set to the right of center.

Left-hand mitered corner. For a coped inside corner, see p. 131.

Right-hand inside corner

To cut a right-hand piece, place it to the left of the blade, with the bottom edge against the fence. Miter angle is set to the left of center.

Right-hand mitered corner. For a coped inside corner, see p. 131.

SPLICING CROWN ON THE WALL

1 Cut a thirty degree bevel on the first piece of crown molding.

To splice lengths of crown molding on the wall, start by laying the crown in position on the saw table and cutting a thirty-degree mitered angle on the first piece ❶. Carefully keeping the bottom edge aligned to the premarked line, install the first piece on the wall with 2½-in. nails driven into the top wall plate framing. Leave the last several feet of the first piece loose until the abutting piece is installed so you can position the joint as needed. Determine the length needed for the second piece ❷ and cut the matching angle on it. Now hold the second piece in place on the line and check to see if the joint fits properly. When satisfied with the fit, apply glue to both faces of the miter joint ❸ and nail the pieces to the wall ❹. Sand the mating surfaces flush after the glue has dried ❺.

2 Measure the length of adjoining piece.

3 Apply glue to the joint.

4 Hold the joint tight and nail in place.

5 Sand the mating surfaces flush after the glue dries.

SPLICING CROWN USING BACKER

1 Cut a matching thirty degree angle on ends of the crown pieces.

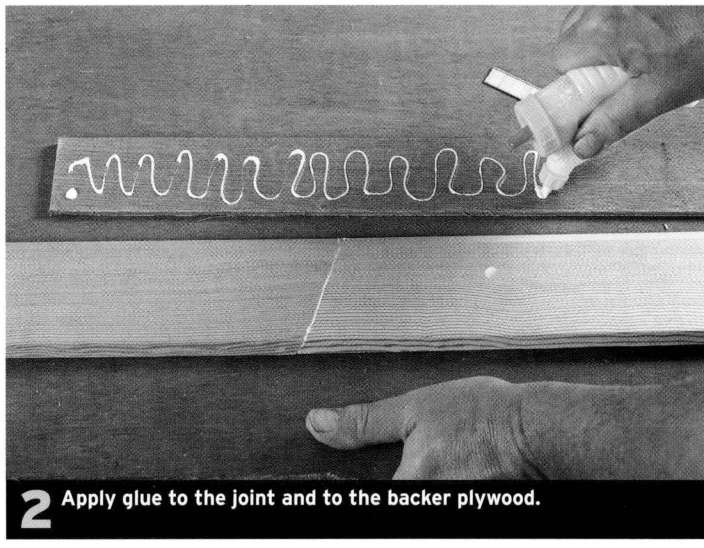

2 Apply glue to the joint and to the backer plywood.

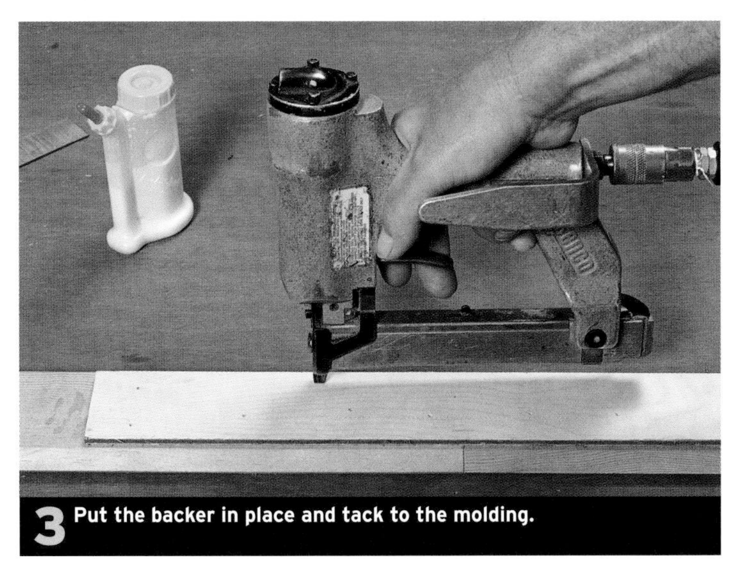

3 Put the backer in place and tack to the molding.

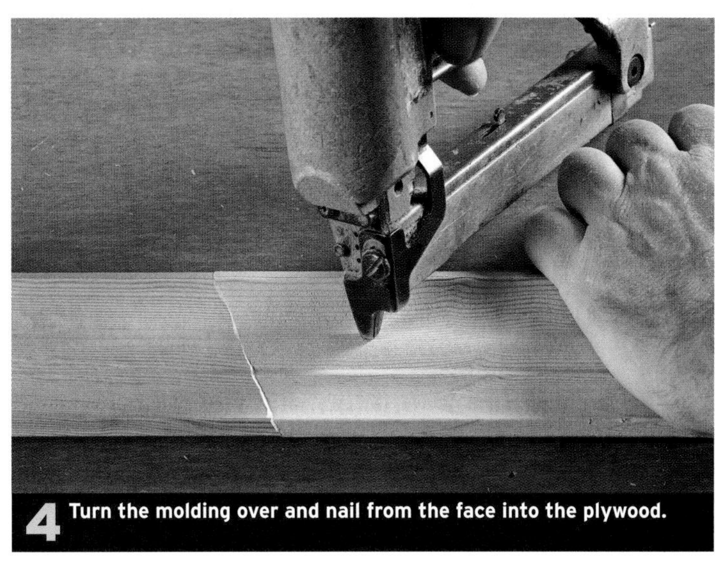

4 Turn the molding over and nail from the face into the plywood.

An alternative to splicing in place on the wall is to join the pieces prior to installation by gluing and nailing a plywood backer to the backside of the crown. On tall ceilings this can save a lot of trips up and down the ladder. Start by cutting the matching thirty-degree angles on the ends of the crown being joined ❶. Rip a piece of 1/4 in. plywood for the backer. Make sure the plywood is not so wide that it will interfere with the crown setting in position correctly. Apply glue to the joint and to the plywood ❷. Using short nails or staples tack the backing piece to the crown from the backside ❸. Then flip the assembly over and make sure that the combined pieces are straight and that the joint is tight. Then nail into the backer from the front ❹. Wait until the glue sets up before moving the piece.

INSTALLING WALL CAP

Wall caps are used to cover up the rough framing on top of a half wall, substituting finished wood for Sheetrock to create a durable and attractive horizontal platform on top of the wall. The two common ways of installing wall caps are: to hang the cap over the edge and apply the trim under the cap or to stop the cap flush to the edge of the wall and then install the trim to the side of the cap, covering the gap while creating a reveal shadow line.

Both types of cap boards have similar installation procedures. On the flush style, you make the cap board to the same width or slightly wider (1⁄8 in.) than the width of the wall. On the overhanging style, you make the cap board to the width of the wall plus the combined thickness of the sub trim and the overhang on both sides. When overhanging the cap, the top of the sub trim is back beveled slightly to ensure a tight fit to the underside of the cap board. Use biscuits and glue or pocket screws on the cap board joints. Shim the top of the wall level, if necessary, and use construction adhesive under the boards.

Before installing wall cap check to see if the top of the wall framing is level. If not, you will need to shim the cap. Cut all the cap boards slightly longer than needed to allow for trimming to fit. Center the first cap board (board A) over the half wall and scribe and cut the end that butts into the wall ❶. Now center this board, and the intersecting corner board (board B), over the top of the half walls butting B up to A. Mark the inside corner intersection point where board B meets board A of the two boards onto the edge of A. Also transfer the location of the outside corner on the Sheetrocked wall to the bottom of board A. Using a straightedge connect these two points with a line and cut the joint on A using the slide saw ❷. Reposition the boards centering them on top of the half walls with the cut miter on board A positioned on top of board B. Using a piece of scrap set under the wall end of board A to hold it level, scribe the miter joint from A onto B and cut the miter on board B. Reposition the two boards on top of the half wall to check the fit. Once the corner miters have been fitted mark board B to length where the cap starts to descend the stair wall ❸. Determine the angle of the miter ❹ and cut the piece to length ❺. Now go back to the corner miter joint and cut biscuit joints in it ❻. After gluing and biscuit joining the miter joint, apply construction adhesive to the top of the wall framing and nail the first two cap boards in place ❼. Mark the bottom of the stair rake wall cap to length by butting the matching miters on the top and marking the bottom of the board where the stair wall ends ❽. Add enough length at the end to provide the same overhang as on the sides and cut the end perpendicular to the floor. Install the sub trim under the wall cap to the wall framing butting it tight to the bottom side of the cap. Cut a slight underbevel on the top of the sub trim to ensure a tight fit to the bottom of the wall caps. At the bottom of the stair, scribe the sub trim to length and cut a compound miter allowing the trim to continue around the corner ❾. Miter and install the bottom piece and proceed around the corner up the other side ❿ to finish the installation ⓫.

1 Scribe the end of the wall cap to the wall.

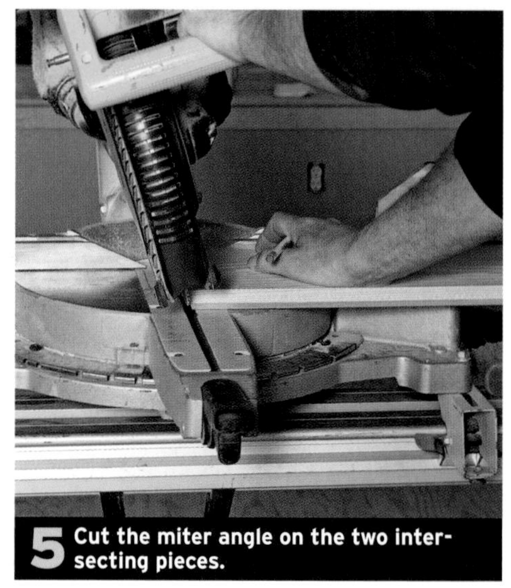

5 Cut the miter angle on the two intersecting pieces.

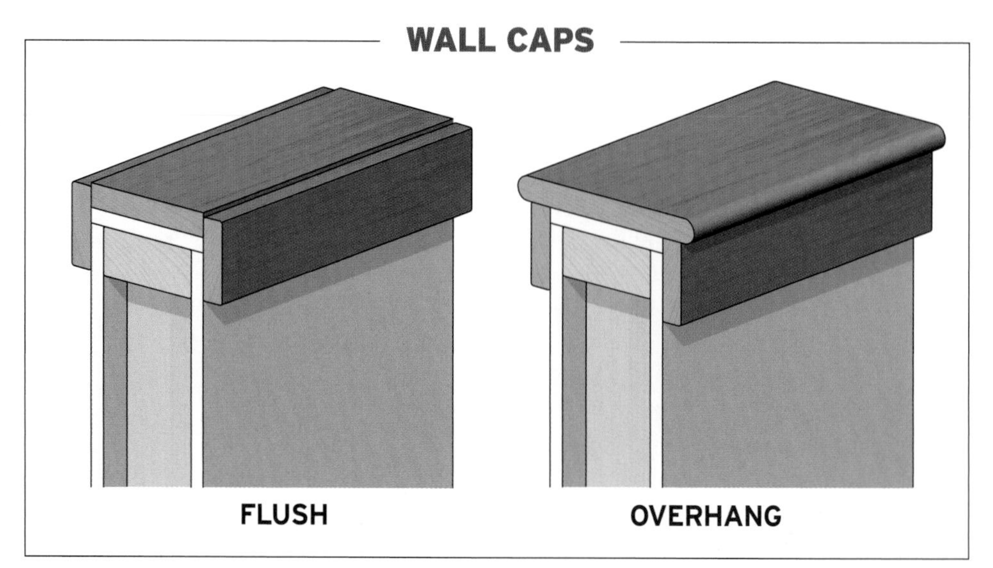

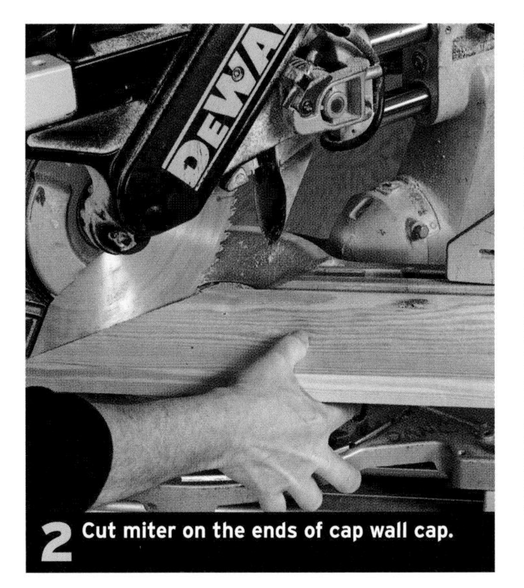
2 Cut miter on the ends of cap wall cap.

3 Mark the end of the second piece of wall cap to length.

4 Determine the angle of the miter cut at the top of the stairs.

6 Cut biscuit joints into the mitered pieces at the corner.

7 Glue and nail wall caps around corner.

8 Mark stair wall cap to length.

9 Cut a compound miter on end of sub trim.

10 Install mitered bottom trim.

11 The finished installation.

MITERED END CAP

1 Draw cut lines onto face of board.

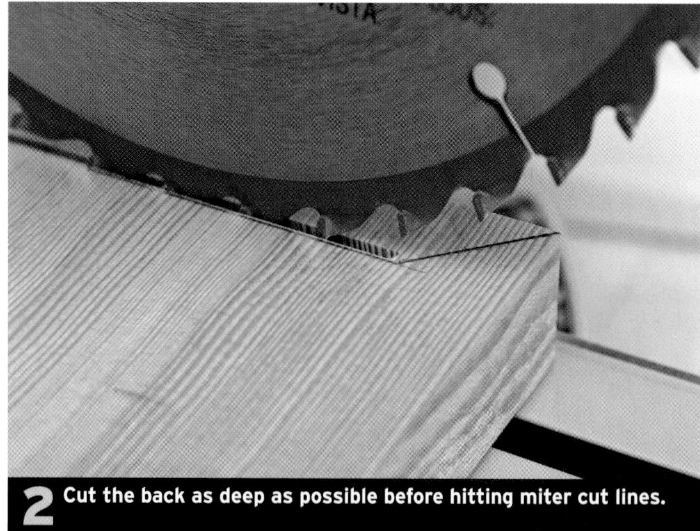
2 Cut the back as deep as possible before hitting miter cut lines.

4 Finish cutting miters with inverted saber saw.

or

Alternately cut miters with handsaw.

6 Put wall cap face down and scribe miters onto end cap stock.

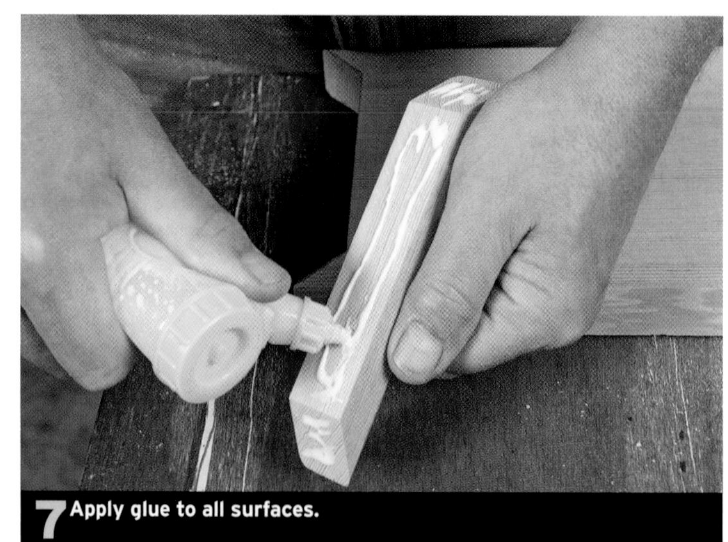
7 Apply glue to all surfaces.

3 Cut the miters before intersecting the back cut line.

When the wall cap is finished natural and installed with its edges hanging out past the sub trim, it is sometimes desirable to eliminate the end grain showing when the cap board ends. The best way to do this is to remove a narrow piece from the end of the board using mitered cuts from each corner and replace it with a piece ripped from the side of another board. This is only necessary on natural finished materials.

To create a mitered end piece on the cap board, begin by drawing out the cut lines for the miter on the face of the board. Using a combination square draw a line across the end of the board 3/4-in. from the end. Then draw a 45-degree line from each corner to the first line ❶. Using a slide miter saw cut as deep as possible on each cut line, stopping the cut at the intersection of the lines ❷, ❸. Though these cuts are incomplete, they will help guide a saber saw or a backsaw in finishing the miter cut ❹. To reduce splintering and to provide a clear view of the cut, invert the saber saw and finish cutting away the waste ❺.

After removing the waste from the end of the cap rip a 3/4-in. wide strip of cap stock on the table saw to serve as the end piece. Set the wall cap board face down on top of the strip and align its edge to the cut line. Scribe the miter joints onto the end piece ❻. Cut the miters on the strip and check the fit. Most likely they will need to be trimmed a bit to get a tight fit. It may also help to cut a slight underbevel on the edge of the cap board cuts by tipping the saber saw slightly and shaving a little bit off the bottom portion of the cut. Once the end cap fits tightly apply glue to all the fitted surfaces and nail in place ❼, ❽. The finished mitered cap installed on end of board ❾.

5 Continue cutting back of end cap and remove waste.

8 Nail end cap in place.

9 Mitered cap installed on end of board.

INSTALLING PERIMETER BAND BOARDS

Band boards add a strong horizontal element to a room, making a high-ceiling room feel visually more comfortable to the eye. Naturally finished band boards fit well in contemporary styled homes. In Craftsman-styled homes the band can become the head trim of the windows and doors. When installing natural finished band boards—and especially when having to splice pieces to cover a long wall—pay careful attention to the color and grain match of the adjoining boards. Start band board installation by marking the height the board is to be applied at and transfer the height around the perimeter of the room using either a spirit or laser level. If the band board will be applied close to the tops of windows or along a soffit, the height will need to be very consistent relative to them to prevent an irregular appearance. If the window heights end up a bit irregular, average the gap between them and the band board out as well as possible and fudge the heights of the band.

On outside mitered corners measure or scribe the first piece to length, then cut the miter and tack in place. Then holding the second corner board in position tight to the end of the first board, scribe the face of the first board onto the backside of the second board. This marks the cut line for the miter on the second board ❶. After the joint fits correctly apply glue to the miters ❷, and nail the corners together ❸. Always join boards over a stud. Find the joint location before cutting pieces to rough length. To splice the pieces, use a 30 degree scarf joint ❹. When measuring a closure piece between two finished boards, use a pinch stick ❺. When installing a closure piece between two finished boards cut the piece long and alternately holding each end in position tight to the corner check the fit of the butt joint ❻. If necessary, trim the angle to fit and cut a 3-degree back bevel on the ends before installing to ensure a tight fit to the face of the board ❼. Tighten any gaps in the joints by inserting a shim behind the back of the board in the corner ❽. The finished perimeter band ❾.

1 Scribe outside face of miter to intersecting corner board.

4 Use a 30-degree scarf joint to splice boards.

RAIL MOLDINGS

Installation of picture and chair rails is similar to that of baseboard and band boards. With the exception of some bolection moldings (transition moldings cut and installed at an angle), these rail moldings lie flat on the wall and use the same type of joints and installation techniques.

Chair rail is traditionally held at about 36 in. off the floor to prevent damage to the wall from the backs of chairs—although that height can vary greatly with style as chair rail also serves as a visual transition between the wall surface and wainscoting. Picture and plate rails are installed between the chair rail and the ceiling while band boards usually run above the tops of the windows and, in the Craftsman style, become the header trim of the windows and doors. Inside corners joints are coped, butted, or mitered and outside corners are mitered.

7 Cut a 3-degree back bevel on board ends.

2 Apply glue to outside miter joint.

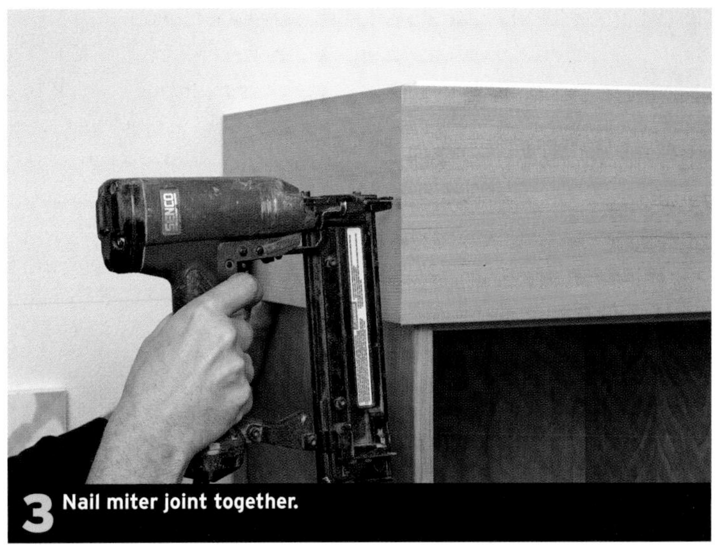

3 Nail miter joint together.

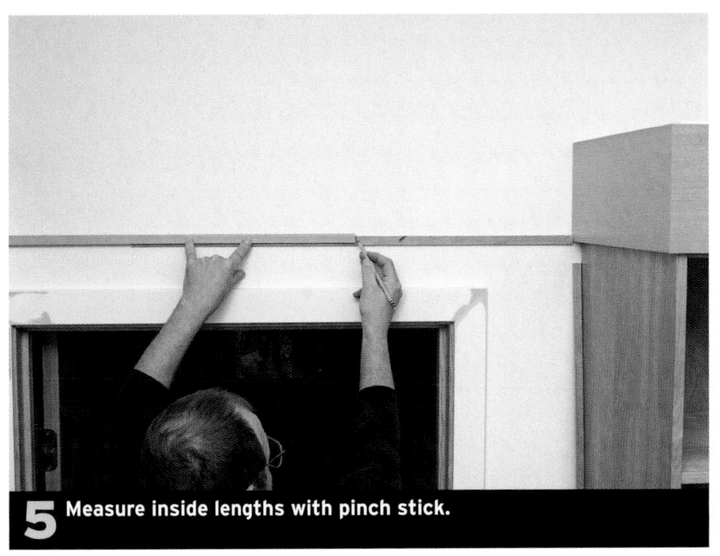

5 Measure inside lengths with pinch stick.

6 Check the fit of the end cuts.

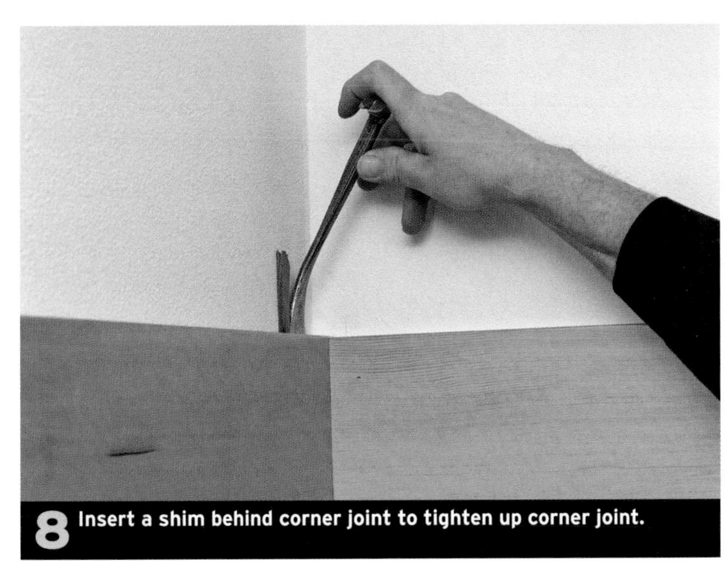

8 Insert a shim behind corner joint to tighten up corner joint.

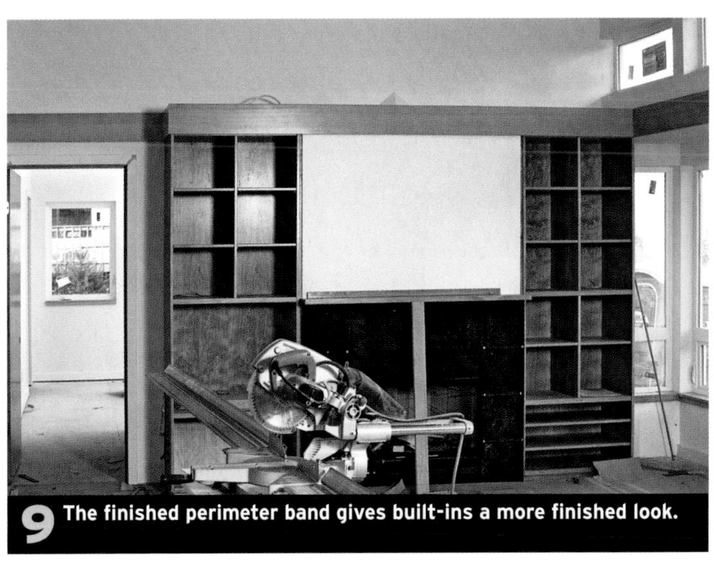

9 The finished perimeter band gives built-ins a more finished look.

WAINSCOT

WAINSCOTING DRESSES UP THE LOWER section of a wall, lending visual interest with complex intersections of lines, shadows, and geometric shapes.

Wether you go with traditional painted wainscot or more casual, naturally finished wainscot, you'll find that both heavily impact and change the character of a room. The average height of wainscot is around 36 in. but I have installed it up to 60 in. high in a Craftsman-style home.

Wainscot capped with a simple chair rail makes a more abrupt change than using a bolection molding, which makes a smooth transition into the wall above. If the panel goes higher up the wall, it's often capped with a decorative plate rail.

This section shows how to build and install both vertical board (tongue-and-groove) wainscot and frame-and-panel wainscot. You'll also find three ways to build frame and panel wainscoting.

Photo by Carolyn L. Bates, www.carolynbates.com

WAINSCOT

Photo by Carolyn L. Bates, www.carolynbates.com

A high wainscot can also be capped by a plate rail. This wainscot was constructed by applying battens over paneling.

Photo by Brian Vanden Brink

Two basic styles of wainscot

There are two basic wainscot styles: a row of solid, vertical boards; and a frame-and-panel. Solid boards are the easiest to install and present a more informal appearance. Frame-and-panel wainscot offers a more formal appearance and requires more planning and effort to install. With this style it is especially important to take electrical outlet and heater locations into account when designing the layout as well as any windows that may protrude down into the wainscot. You want to avoid having these elements cut partway into a rail or stile.

Board wainscot is built up of individual tongue-and-groove boards run directly from the floor to the chair rail or, more common, set between a baseboard (which serves as a bottom rail) and a top rail. A common option, which I show here, is to run the boards behind a rabbet cut in the rails to hide any irregularities in the butt joints. The boards are glued and hidden-nailed through the tongues to the wall. This secures the boards until the construction adhesive sets up. Though solid boards are subject to shrinkage, typical wainscot boards are narrow enough to minimize the amount.

Photo by Brian Vanden Brink

This bead board wainscot in an entry foyer is higher than usual, but the height is perfect to create a place to hang coats from the top rail.

Use battens to create dimension over a painted wall. Or use wallpaper or fabric in the sections between battens for an entirely different look.

Photo by Charles Miller, courtesy *Fine Homebuilding*

This elegant frame-and-panel wainscot takes careful layout and precise cutting to build, but the results are dramatic.

Photo by Carolyn L. Bates, www.carolynbates.com

Wainscot and chair rail can be painted in contrasting colors to create different effects.

CHOOSING MATERIALS

The materials used for wainscot range from a variety of paint-grade materials to high-grade, naturally finished lumber and plywood. (Some plywood products feature a pattern that mimics solid wood boards.) Paint-grade materials are MDF, lower grades of plywood, and softwood lumber. The choices are endless for naturally finished materials, limited only by your budget and imagination.

MAKING RABBETED BASEBOARDS

1 Set an adjustable square to the thickness of the wainscot board.

2 Transfer the board thickness to the baseboard.

3 Set tablesaw fence so blade cuts on waste side of line.

4 Set saw fence and blade height to remove waste from rabbet.

A rabbeted baseboard and top rail hide any gaps caused by irregularities in the end cuts of the wainscot boards. To lay out the rabbet, set an adjustable square to the thickness of the wainscot board and transfer this dimension to the end of a piece of baseboard stock ❶, ❷. Set the tablesaw fence so the saw blade lines up on the waste side of the mark on the end of the board and set the blade depth arbitrarily from 1/4 in. to 1/2 in. for the first cut ❸. This depth determines how far the boards will fit into the rabbet. Make the cut then reset the fence so the blade lines up with the end of the previous cut. Adjust the depth of the cut to remove the waste ❹. Use a featherboard to keep the stock tight to the fence.

Wainscot height

How high should your wainscot be? There is no standard, but there are some rules of thumb, depending on the purpose of the wainscot and the height of the room. Chair rail capped wainscot is usually between 32 in. to 36 in. The higher the ceiling, the higher the wainscot to remain in proportion, but avoid ending the wainscot at the midpoint of the wall. For wainscot capped with a plate rail, the wainscot can be higher. In some interior styles, such as Arts and Crafts, high wainscot 60 in. or more is the norm.

INSTALLING RABBETED BASEBOARDS

1 Butt two pieces of base together and mark inside face of rabbet.

2 Mark the cut lines and remove the waste with a backsaw.

3 Installed base ready for wainscot.

4 Cut and glue corner transition piece to baseboard.

Rabbeted base is installed in the same way as regular baseboard with the exception of inside corners and around radiused outside corners.

See the section on Baseboard Molding, pp. 116–127.

Inside corners

When running rabbeted base into an inside corner, you must make a notch in the side of the rabbet on one piece to allow the wainscot boards to run all the way into the corner. First cut the intersecting pieces of base to length and then butt them together in the corner. Now trace the inside edge of the rabbet on the face of the second board ❶. Use a backsaw to remove the waste from the corner ❷. Nail the board running all the way to the wall in first. Before nailing the second abutting board in place, be sure that its end sits tight against the face of the installed board ❸. Scribe and trim as necessary until it does.

Radiused outside corners

When you are dealing with irregular plaster walls (as shown here), fitting trim around a corner becomes more a matter of trial and error rather than working to direct measurements. It may also require fastening with glue rather than nails. Begin by cutting the small corner transition piece. On a three-piece corner all the pieces are mitered at $22\frac{1}{2}$ degrees. If you are dealing with a typical one-inch radius corner, cut the transition piece $\frac{5}{8}$ in. long measuring between the short points of the miters. Line the transition piece up equally across the intersecting walls and mark the miters on the floor. Measure to these lines to determine the lengths of the intersecting base pieces.

INSTALLING RABBETED BASEBOARDS (CONTINUED)

➔ **See "Installing Baseboard around Radiused Corners," p. 120.**

To install the trim, glue and pin nail the short corner transition piece to one of the runs of baseboard ❹. Make sure the pins are sized small enough so they won't come through the face of the base. Apply construction adhesive to the wall and set the first two assembled pieces in place ❺. Set the third piece in position with construction adhesive on the wall and yellow glue on the miter ❻, ❼. When you are satisfied with the fit of the joints, pin the miters together ❽. Once the adhesive sets up, the trim will become permanently locked to the wall.

5 **Apply construction adhesive to wall.**

6 **Glue miter of closing piece of base.**

7 **Adjust pieces for final fit.**

8 **Nail corners transition piece to base with pin nailer.**

INSTALLING TONGUE-AND-GROOVE BOARDS

1 Cut out drywall to make room for backer boards.

2 Screw backer boards into studs.

3 Cut a 45 degree miter on the groove edge.

4 Nail the board on one side of the corner.

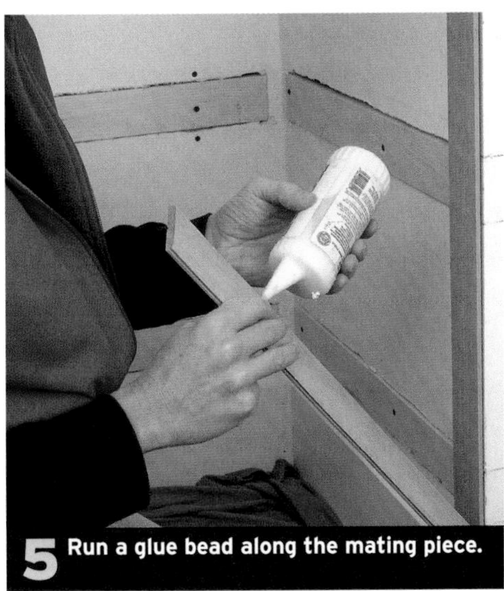

5 Run a glue bead along the mating piece.

6 Holding the second board tight, nail it down.

An alternate to using adhesive and rabbeted running moldings is to let in solid board nailers. You may need to retrofit nailers by cutting away the drywall. Snap layout lines around the room and then saw out the drywall ❶. Screw backer strips into each stud ❷. Begin with the outside corners where you will use a full width board. Cut a 45 degree miter along the groove edge on the tablesaw on each of the two corner boards ❸. Nail the board to one side of the corner holding it as close to plumb as possible ❹. Run a bead of glue down the mating piece ❺. Tack the second board into the wall, and then nail through the miter to pull it tight ❻. For each successive board, nail only the tongue side of the board. Inside corners will butt against one another. Plan to have the inside boards about the same width as the outside boards.

LAYING OUT BOARD WAINSCOT

When planning the layout of wainscot boards, start at the most visual spot and use a full board (ripping off a slight bevel on the edge meeting the wall) to start with. On an outside corner plan full width boards for appearance sake. On an inside corner the boards don't have to be full width but the corner will look better if both boards coming into the corner have the same apparent width. If the wainscot run ends into a wall that is out of plumb, the boards can be spaced slightly further apart at the top or bottom to make up the difference, or they can be scribed and cut to fit if the gap is too great or the surface irregular.

INSTALLING AROUND A RADIUSED CORNER

1 Mark wainscot for width.

2 Rip a 22½-degree angle on both sides of corner piece.

3 Apply glue to mitered corner.

4 Nail corner transition piece into adjacent board with pin nailer.

Make the pieces of wainscot on either side of the corner piece the same width. Start by setting the wainscot board meeting the corner on top of the baseboard and butt it to the last installed piece. Then mark it for width ❶. Use the tablesaw to rip a 22½-degree miter on the marked end. Now measure and rip to width the narrow transition piece using a 22½-degree angle on both faces ❷. Apply yellow glue to the miter and pin nail it to the previous piece ❸, ❹. Finish the corner by gluing and nailing the next mitered corner piece to the transition piece ❺.

5 Installed base and wainscot around corner.

INSTALLING A CLOSING BOARD TO AN IRREGULAR WALL

Wainscot runs will usually end at another finished panel or butt into a standing molding. When, however, they intersect an irregularly finished plastered or masonry wall they will require scribing and cutting the closure board to fit.

Prepare the closing board by ripping off the back half of the groove so you can install it without having to engage the tongue ❶. Set the scriber to the widest dimension that the board will have to fill ❷. Mark that distance on the board by setting the point of the scriber on the edge of the groove ❸. Using a level to keep the board plumb, touch the board to the wall and reset the scriber to the spread between the wall and the mark that you just made on the board. Now run the scribe along the length of the board, transferring the wall profile to the board ❹. Cut to the line using an inverted saber saw, tilting the saw slightly to produce an underbevel. Clean up the cut with a belt sander or block plane ❺, ❻, then face nail the board into the corner framing stud ❼.

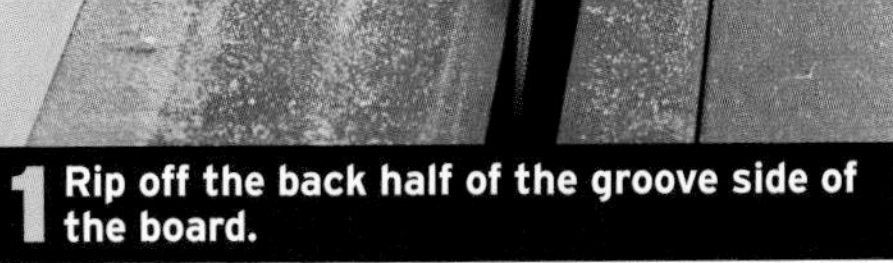

1 Rip off the back half of the groove side of the board.

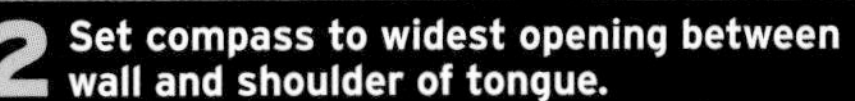

2 Set compass to widest opening between wall and shoulder of tongue.

3 Set the compass point on the edge of closing board and mark width.

4 Plumb the board next to wall, reset compass, and scribe the line.

5 Use the inverted saber saw to cut to the line.

6 Use a belt sander to create undercut and trim to the line.

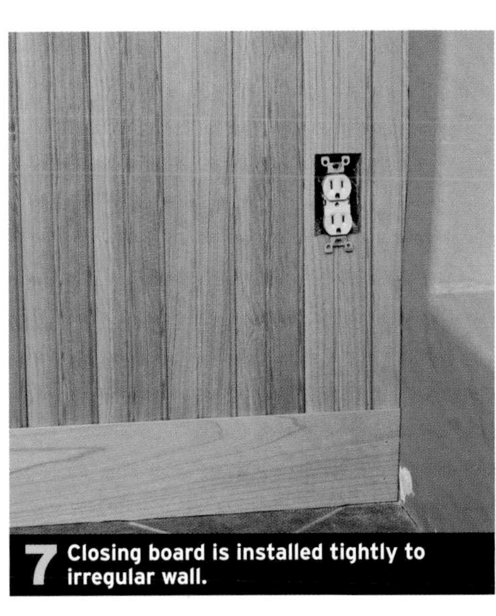

7 Closing board is installed tightly to irregular wall.

MARKING AND CUTTING FOR OUTLETS

Set the first board that must be cut around the outlet in place and mark the top and bottom edges of the outlet box along one edge ❶. Hold a story stick to the base of the last installed board's tongue and mark the two vertical sides of the outlet box ❷. Transfer the measurements from the stick to the board at the top and bottom marks ❸. Draw in the outline of the box to the marks with a combination square ❹. Now drill a hole large enough for a saber saw blade to fit through in two corners ❺. Use an inverted saber saw to cut out the ❻. First making sure the power is turned off at the box, slip the outlet through the hole as you set the board against the wall. Secure the board and reinstall the outlet with the tabs now bearing on the face of the wainscoting ❼.

1 With board next to outlet mark top and bottom of outlet box.

2 Mark outlet sides on short story "stick."

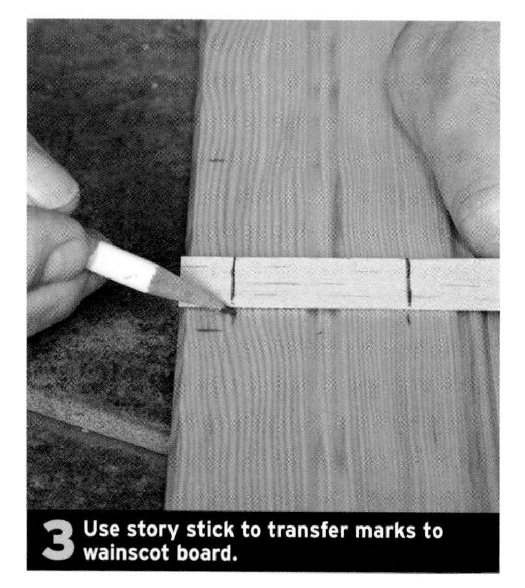

3 Use story stick to transfer marks to wainscot board.

4 Use a combination square to draw outline of outlet.

5 Drill starting holes for the saber saw blade.

6 Cut the waste out with inverted saber saw.

7 Install the outlet after fastening the board to the wall.

INSTALLING WAINSCOT CAP

The cap is installed on top of the wainscoting and screwed or nailed to the wall studs or to the top of the wainscot. The outside and inside corners are mitered using glue and nails or screws to fasten the miters together. The railing must fit tight to the top of the wainscot to eliminate any gaps or if the chair rail is wide enough a transition molding may be used under it, which will cover up any gaps.

In many cases the cap will need to be scribed to fit tight to the wall. For this reason, rip the chair rail stock wider to provide the desired overhang after the scribed material is removed. Start by cutting the cap to length. Set it in place and scribe it to the wall by adjusting the scribe to span the largest gap between the wall and the chair rail. Add 1⁄8 in. to that setting to ensure that the scribe won't slip off the edge of the chair rail while you run the scribe ❶. Use an inverted saber saw to cut to the scribe line, trimming as necessary with a belt sander or block plane ❷. Once the cap fits tightly to the wall, fasten it in place. Drill a countersink and pilot hole at each stud location and screw through the edge of the cap and into the wall studs, drawing the cap tight to the wall ❸, ❹. Now brush a bit of yellow glue into the counter sink holes and then lightly tap in a plug ❺. After the glue has dried trim the plug with a sharp chisel and then sand it smooth ❻, ❼.

1 Scribe chair rail to irregular wall.

2 Cut to scribe line with inverted saber saw.

3 Drill a countersink pilot hole at stud location.

4 Drive screw through cap, Sheetrock, and into stud.

>> >> >>

INSTALLING CAP (CONTINUED)

5 Lightly swab glue on sides of hole and tap in bung.

6 Trim plug flush with sharp chisel.

7 Sand the surface smooth.

INSTALLING RABBETED CAP

Install the rabbeted cap using the same techniques to determine lengths and to go around corners that you used to install the base rail. You won't, of course, have to scribe an edge, as you will simply sit the top rail on the top end of the wainscot boards. Use a stud finder to locate the framing.

Set the rail on top of the wainscot boards ❶ and then nail or screw the rail in place into the studs ❷.

1 Set rail over the wainscot boards.

2 Nail the cap to the wall framing.

INSTALLING CAP AT 90-DEGREE CORNERS

Use a proportional divider or angle finder to determine the degree setting of the outside corner ❶. Because the build up of joint compound makes it difficult to obtain a true angle reading for the rail corner, be sure to cut and test a pair of scrap pieces of rail cap before cutting the rails themselves. Trim the miter cuts as necessary to achieve a tight joint ❷. Then glue and screw the corners together ❸. Fit an inside 90-degree corner following essentially the same procedure: Use a proportional divider or angle finder to determine the angle of the miter cut ❹. Cut the first run of cap at that angle and install it. Then cut the same miter on a piece of scrap and test the fit. If it does, use that angle to cut the end of the intersecting run ❺.

1 Determine corner angle with proportional divider.

2 Check corner angles using test pieces and recut if necessary.

3 Glue and screw corner together.

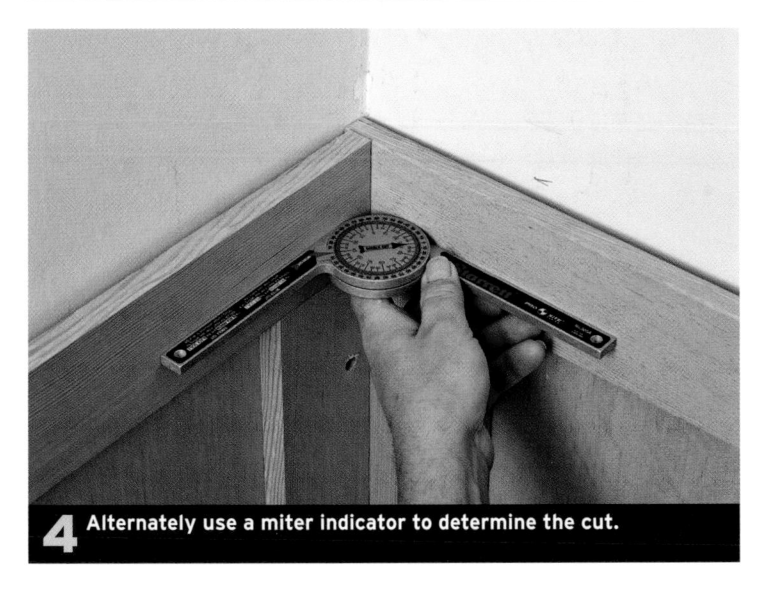

4 Alternately use a miter indicator to determine the cut.

5 Once the angle has been determined cut and install the pieces.

INSTALLING CAP ON RADIUSED OUTSIDE CORNERS

Installing chair rail around a radiused corner adds a bit of visual punch to an otherwise conventional 90-degree mitered corner. The transition piece softens the corner eliminating the right angle and combining with the inside scribed radius to create a very custom look. Start by determining the length of the first wall piece by using the inside corner of the miter on the wainscot below as a reference point ❶. If you have trouble setting the end of the tape exactly on the inside corner, set the 1 in. mark on the tape onto the corner and deduct an inch at the other end. Cut a 22½-degree miter on the end of the run leading up to the corner and screw it to the wall, positioning it directly over the edge of the miter. Cut another 22½-degree miter on a piece of rail stock and, while holding it in position tight to the corner and to the miter of the installed piece, mark the outside profile of the wainscot corner onto the underside of the stock ❷. Again at a 22½-degree angle, cut through the intersection point of the wainscot corner marked on the bottom side of the stock ❸. Before you can install the transitional piece to the miter, however, it is necessary to shape the back to the corner radius. Hold the corner piece in place and use a pencil scribe to transfer the radius onto the piece ❹. Use a belt sander (or alternately a drill mounted sanding drum) to cut to the scribe line ❺, ❻. Now test fit the angle on the second wall piece to the small transition piece, trimming the angle on the second piece if necessary. After everything fits, glue and nail the corner together ❼.

1 Cut the first leg by measuring to inside corner of miter for length.

4 Scribe radius on inside of corner piece.

OTHER RUNNING-MOLDING PROFILES

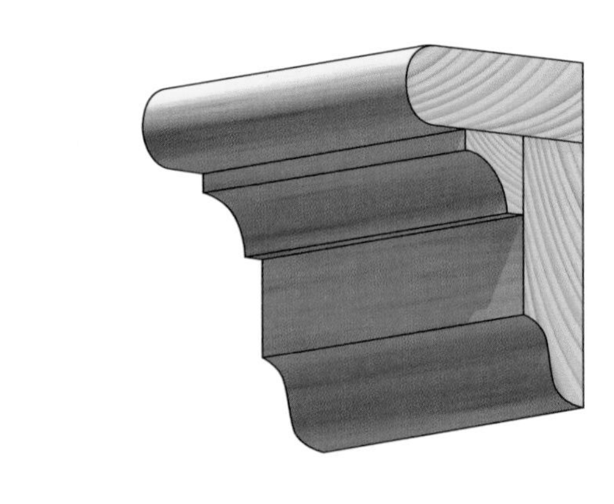

CHAIR RAIL
Usually shallow with a wider top to protect walls from chair backs.

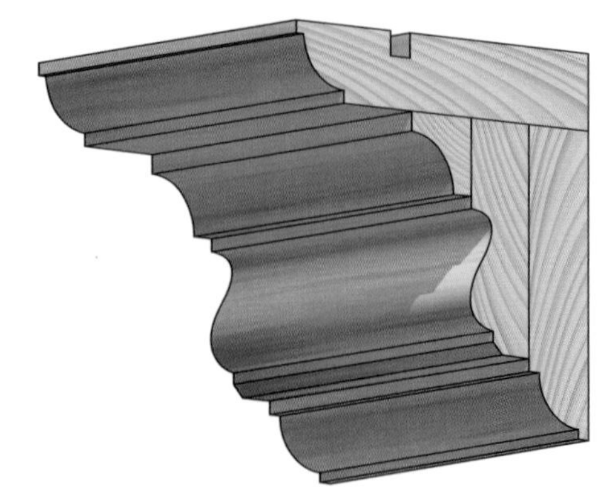

PLATE RAIL
The deeper profile allows for a shelf with a groove for plates.

2 Mark the corner intersection onto bottom of stock.

3 Cut the corner piece to length.

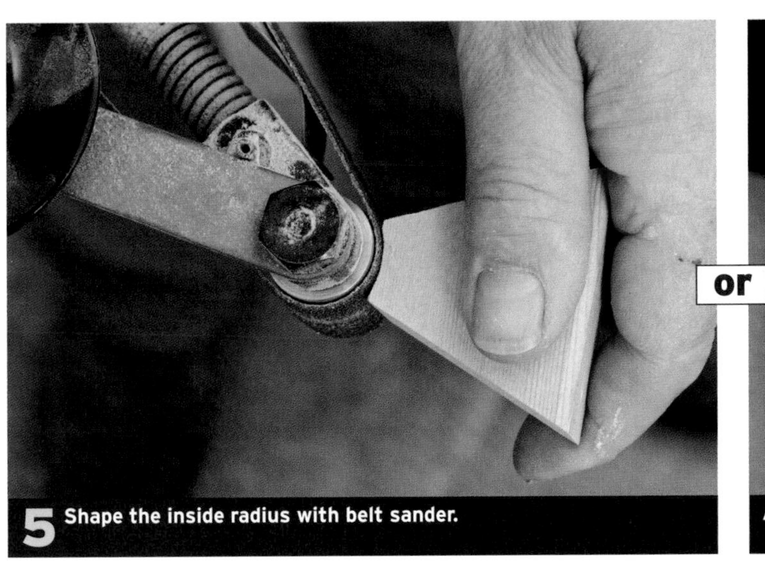
5 Shape the inside radius with belt sander.

or

Alternately use sanding drum to shape inside radius.

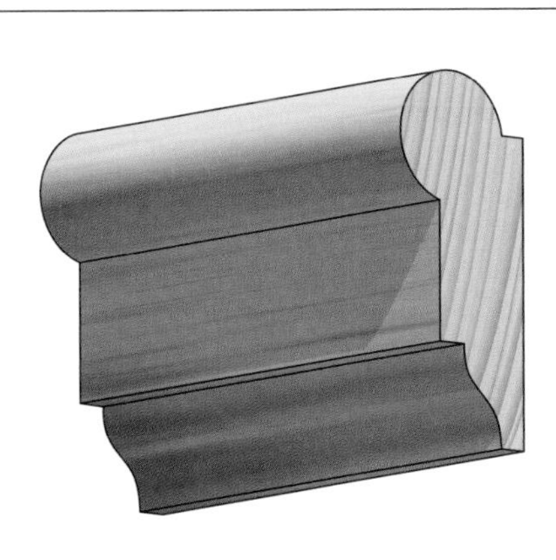

PICTURE MOLDING
A thin profile with a slot or curved lip to hold picture hooks.

6 Nail completed corner together.

LAYING OUT FRAME-AND-PANEL WAINSCOT

First, make a scale drawing of the wall, being sure to include all electrical outlets, doors, and windows. Then rough in different panel layouts, working around the wall utilities to keep them out of the frame and entirely within the confines of the panel and letting your eye tell you what lengths and heights look best.

Avoid having outlets land on, or partially into, the stiles. If this is unavoidable for proper proportioning of the frame and panels, the best solution is to move the outlet. The next best solution is to add a piece of stile material to fill in the area around the outlet and use a box extension to bring the box out to the face of the stile material. In some cases you can use a piece of stile material to fill in the area around an outlet.

When planning the overall configuration of the wainscot frame-and-panel units, try to avoid running a unit from an outside corner into an existing panel. Instead, plan the runs that come off outside corners to slide behind intersecting panels at inside corners. This avoids having to scribe the corner stile to fit to an existing panel at one end and to a corner at the other. There will be no choice but to do this, however, when the unit must run into a standing molding (such as a door casing). In this case, make the frame-and-panel unit at least 1/4 in. longer than necessary to account for scribing and trimming it to fit.

Proportioning panels

Much of the time, I find that the panels come out close to the proportions of a golden rectangle (a ratio of five to eight according to an ancient Greek proportioning system based on growth-progressions found throughout nature, including our own bodies).

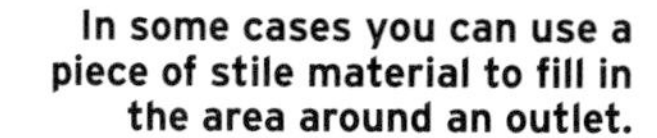

In some cases you can use a piece of stile material to fill in the area around an outlet.

DIMENSIONING FRAME AND PANEL

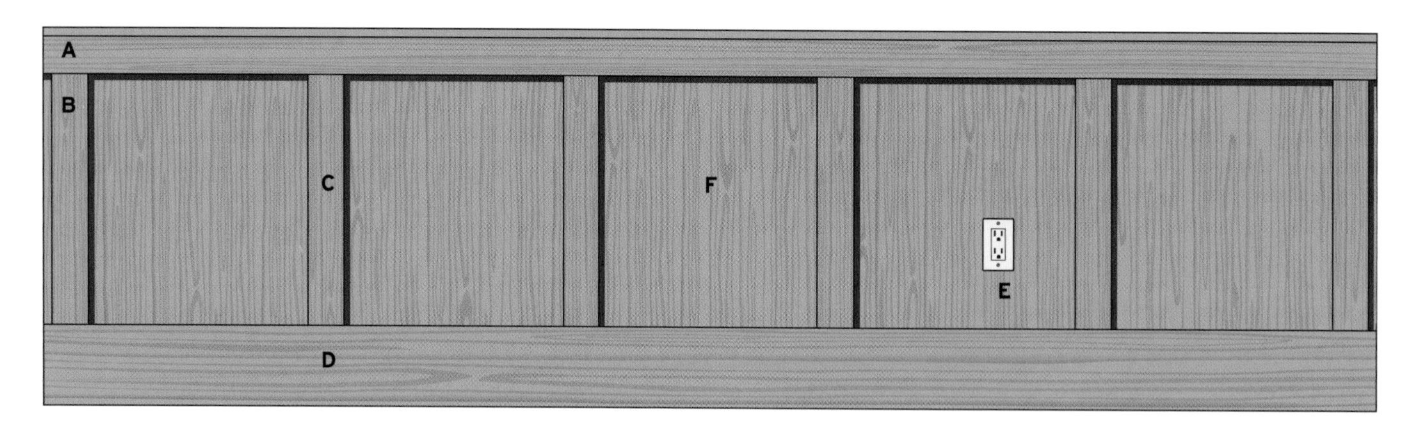

A. TOP RAIL: approximately equal to width of stiles in field (typically 3 in. to 3 1/2 in.).

B. END STILES: where wainscot abuts in corner adjust width to compensate.

C. FIELD STILES: usually same width as top rail.

D. BOTTOM RAIL: several inches wider than top rail to provide visual anchor to floor.

E. OUTLETS: avoid outlets landing on stiles.

F. PANELS: determine spacing by subtracting combined stile widths from overall length of wall. Then divide the result by the number of panel spaces to derive panel width.

Example
Overall length: 120 in.
Field stile width: 3 1/4 in.
End stile width: 4 in.
120 - (4 x 3 1/2) - (2 x 4) = 98
98 ÷ 5 = 19.6 or approximately 19 5/8.

When using plywood for panels, I save considerable money by laying out the panels to minimize waste. I do this by experimenting with the layout so it will utilize as much of the plywood's width as possible. For example, I try to make panels just under 24 in. or under 16 in. so I can use the offcuts to produce more panels from the 48-in.-wide sheet. Also when deciding on the widths of the stiles and rails, I use standard lumber sizes as much as possible.

Proportioning rails and stiles

I usually size the stiles and the top rail close to the same width. I make the bottom rail wider to anchor the assembly visually to the floor—usually adding two or three inches to its width is adequate. On an inside corner, the intersecting end stiles should be the same apparent width as those in the rest of the frame. This requires that I make the first installed stile wider by the thickness of the intersecting panel.

Plan for outside corner stiles to run full length from the floor to under the chair rail. This eliminates having to deal with small pieces and makes for a more attractive corner. The other stiles run between the top and bottom rails.

EFFICIENT LAYOUT SEQUENCE

The below layout sequence minimizes the amouth of pieces requiring exact fitting by butting into a GWB corner.

6

5

7

1

2

3

4

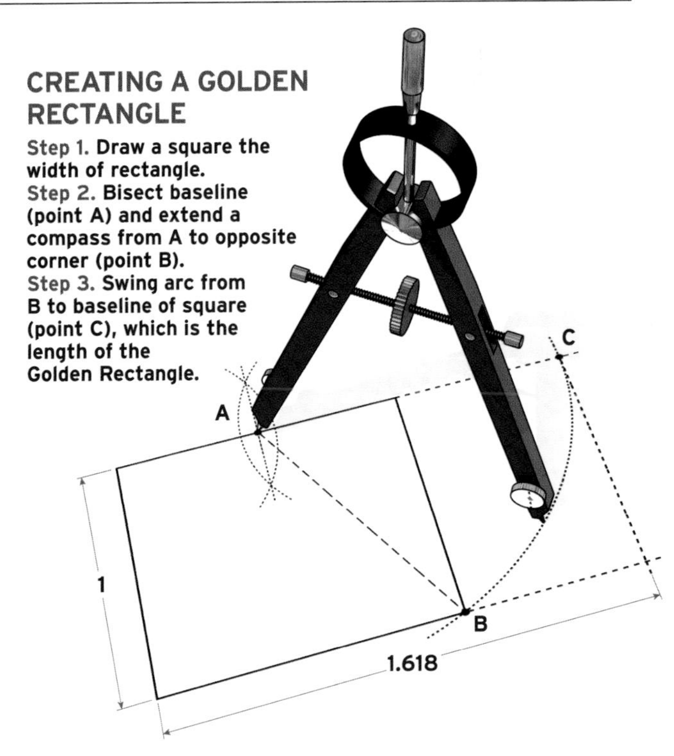

CREATING A GOLDEN RECTANGLE

Step 1. **Draw a square the width of rectangle.**
Step 2. **Bisect baseline (point A) and extend a compass from A to opposite corner (point B).**
Step 3. **Swing arc from B to baseline of square (point C), which is the length of the Golden Rectangle.**

Frame-and-Panel Wainscot Checklist

1. **Approximate size of panels is combined with desired stile width to determine spacing.**
2. **Outlet and frame locations need to be checked beforehand so they do not conflict. Frame layout or outlet location may need to be altered.**
3. **Determine window locations.**
4. **Plan layout sequence for maximum efficiency.**

ASSEMBLING THE FRAME-AND-PANEL UNIT

Using pocket screws easily creates flush joints and saves a lot of time because there is no need to clamp joints and wait for glue to dry. Another advantage is that the frame with pocket screws is the frame that can be checked for fit before the plywood is attached to the back. If the end stiles are not glued when assembled, it is a simple matter to remove them and trim them to fit. If this assembly is going between two paneled walls make sure when installing the other assemblies that they are plumb at the corners and cut the assembly going in a bit long so it can be sprung in place between the two. It is easier to work from only one installed panel assembly leaving the other end short to be covered up by the next panel.

➔ See "Laying Out Frame-and-Panel Wainscot," p. 160–161.

You'll find that It is important to have a flat platform on which to assemble the frame-and-panel units. Although the floor will work it is not very comfortable to work on. I make a platform out of plywood and some 2x4s laid on top of sawhorses.

To produce a square wainscot frame with tight-fitting joints that will square itself as it is assembled, cut all the stiles to the exact same length with accurate 90-degree cuts on the ends. A repetitive cutting fixture on the miter saw makes quick work of this.

TRADE SECRET

When cutting plywood, a piece of solid foam board placed under the sheet will provide a solid base to cut on and support the edges.

Assembling the frame

After determining the general spacing of the rails and panels for the wall section, the next step is to lay out the stiles and panels on a story pole accurately from which you will transfer the stile locations to the top and bottom rail ❶.

To ensure that the panels are of equal length, create a layout stick cut to the estimated width of one panel and one stile. Now mark the width of the first stile at one end of the story pole, place the panel end of the layout stick to that mark, and transfer the location of the next stile (marked on the layout stick) to the story pole. Continue setting the panel end of the layout stick to the outside edge of the stile marked on the story pole and transfer the next set of panel and stile marks, moving down the story pole until you reach the mark representing the opposite wall. If the spacing ends up too long or short, change the stick's length slightly and repeat until it's correct. Once all the pieces are marked for layout and cut to length, position them all on the table front face down. Drill pocket holes for all the joints on the stile ends ❷. After the holes are drilled, carefully line up the stiles to the marks on the rails and clamp and pocket screw them together ❸. After the first several stiles are joined to the rails check the assembly for square ❹.

Attaching the panels

I install the panels to the back of the assembled frame with screws so I can easily adjust the position of the panels if necessary. This also ensures that the panels will always remain tight to the frame regardless of any deviation in the wall surface. This is a great system to use when applying wainscot to irregular or crooked walls because the assembly can span the low spots and at the same time it keeps the panels tight to the frame. I apply a heavy bead of construction adhesive in any low spots in the wall to give the panel solid backing.

Lay the plywood sheet on its face and use a straightedge or Sheetrock square to mark the cut lines ❺. Using a saw guide, cut the plywood from the backside to eliminate any chipping of the face veneer ❻. Now spread the panels out on the back of the frame and drive in a screw every 6 in. along all the edges and down the stiles ❼. If the outlet box hole needs a little adjusting to correctly fit over the outlet box, the panel can be removed and adjusted as necessary to line the hole up over the box. Note that it may be necessary to trim the outside edges of the plywood panel to allow a change of position. Pre-installing the plywood to the assembled frame works well as long as the length of the room does not dictate having to splice the rail sections together, at which point pre-assembly becomes impractical.

TRADE SECRET

The only time I do not pre-assemble the entire wainscot unit is when I cannot get rails long enough to go the entire length of the wall (usually when over sixteen feet.) In those cases it would require splicing of the rails, which would make pre-assembly very difficult. Moving anything over sixteen feet long into place is also difficult.

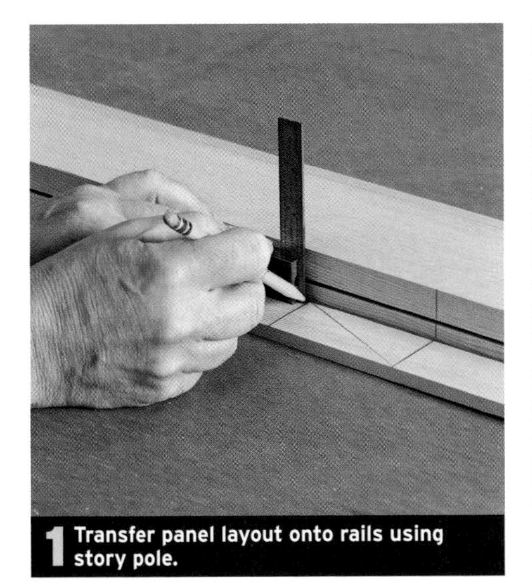

1 Transfer panel layout onto rails using story pole.

2 Drill pocket holes on backside of stile ends.

3 Assemble stiles to rail using pocket screws.

4 Check frame assembly for square.

5 Mark cut lines onto back of plywood.

6 Cut plywood panels to size with circle saw and guide.

7 Screw panels to back of frame.

LOCATING AND CUTTING HOLES FOR OUTLETS

To mark out for wall outlets use a story pole or tape measure. Choose a story pole if there are multiple outlets on a wall as you can mark all the outlets at one time. Use a tape measure when the outlets are spread out on a long wall as story poles become unwieldy. Always use a story stick to locate the top and bottom of the box. First mark the outlet box horizontal locations on the story pole ❶, ❷, then index the pole to the edge of the frame and transfer the marks to the back of the plywood panel ❸. When cutting from the back of the plywood use a non-inverted saber saw to keep the splintering on the backside ❹. Remember that because the panel edges are completely hidden behind the frame the panels may be adjusted a bit if the outlet has been cut a little out of position.

1 Record outlet box locations onto horizontal story pole.

2 Mark top and bottom of outlet box on story stick.

3 Transfer marks from story poles to back of panel and draw outline.

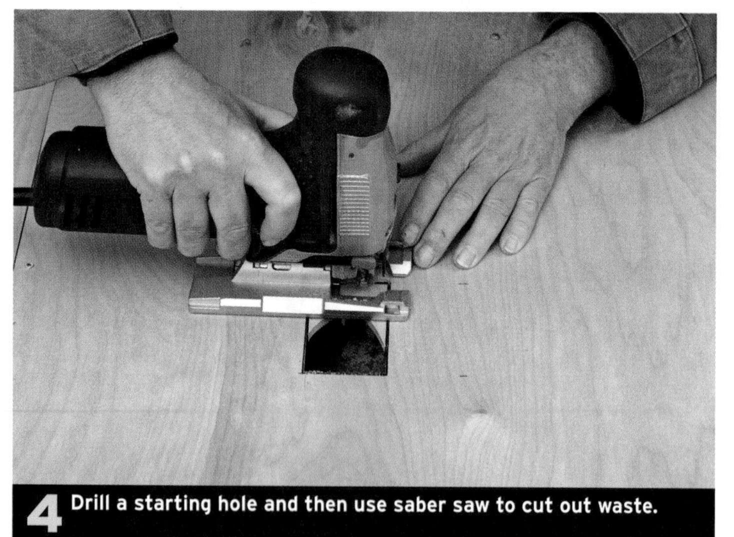

4 Drill a starting hole and then use saber saw to cut out waste.

INSTALLING FRAME-AND-PANEL ASSEMBLY TO WALL

1 Move assembled wainscot unit into place.

2 Level assembly with shims set between bottom rail and floor.

3 Carefully plumb at corners where adjoining wainscot intersects.

After assembling the wainscot frame and panel and cutting in the outlet box holes carry the unit to the wall and set it on spacers (generally 1/2 in. thick if the floor covering is going to be carpet) ❶. Level the assembly by inserting shims between the bottom stile and the 1/2 in. spacer ❷. If there are to be intersecting panels in the corners be especially careful to install the wainscot unit plumb at the corner so that the abutting panel will not need to be scribed to fit ❸. Adjust as necessary with shims or double-threaded adjusting screws.

When you're satisfied with the hanging angle, screw the unit through the rails and stiles to the wall studs. (Screws eliminate nail holes but do require that the holes be plugged.)

WORKING AROUND REGISTERS

Heat or cold air return registers in a wainscoted wall can complicate the process and may determine how the wainscot will be installed either piece by piece or preassembled. If the register is low enough it may be simply cut into the bottom rail leaving enough material to keep the rail intact and the wainscot may be installed preassembled. If the register requires the bottom rail to be cut in half it is best to assemble that wainscot section piece by piece on the wall.

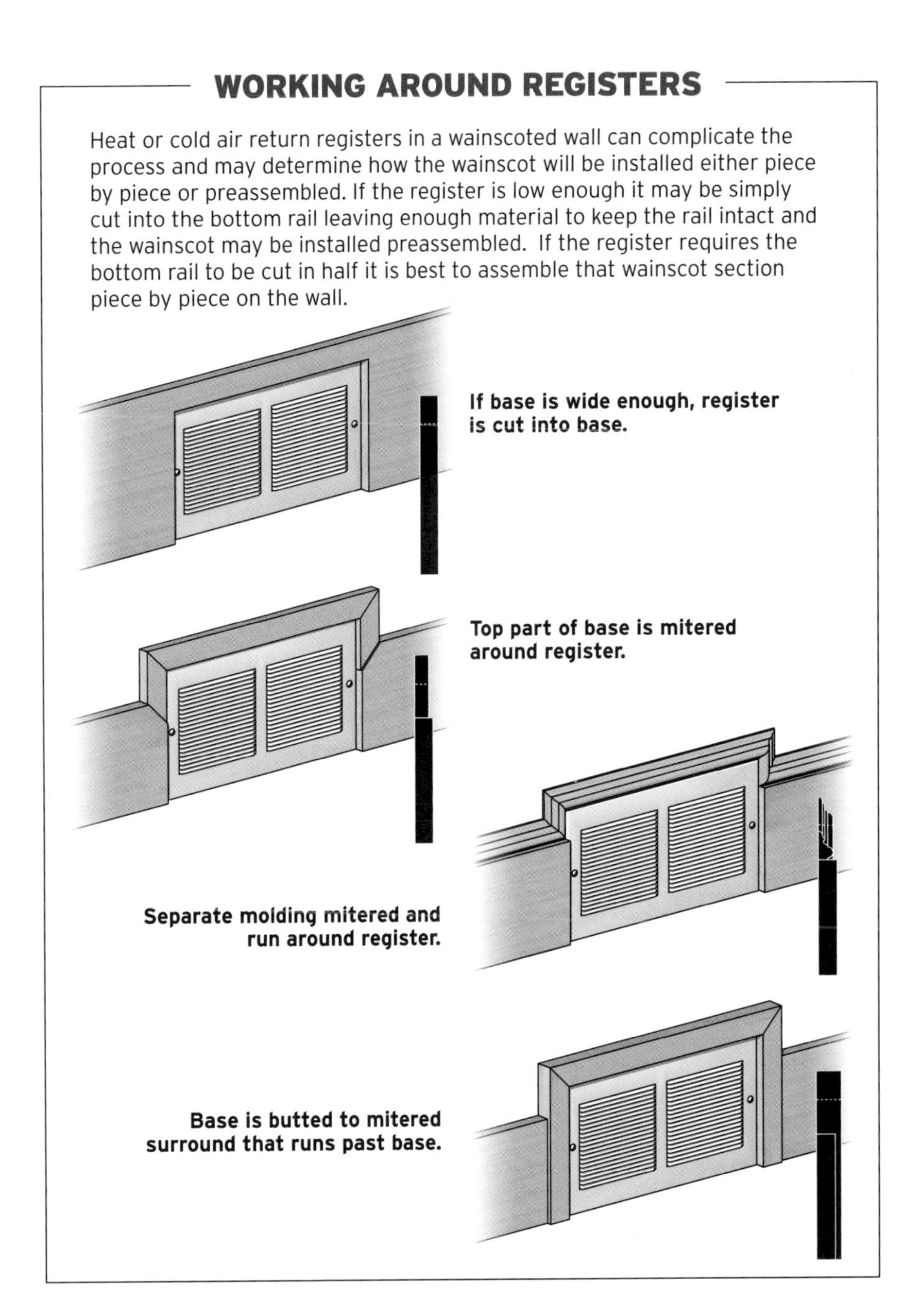

INSTALLING ASSEMBLED PANELS ON INSIDE CORNERS

Determine the corner miter angle before assembling or installing the corner pieces as the corner may not be square. Use either a proportional divider or a miter gauge to do this.

Begin by fastening the first section into place using shims or double-threaded screws at the corner to bring it plumb ❶. There isn't a need to scribe the stile to the wall as the edge will be hidden behind the abutting unit. Plumb the unit carefully, however, to avoid having to scribe the end of the second unit to fit. Now level the second unit with shims and butt it to the first unit. For final fitting, you can choose either to adjust the first panel slightly out of plumb if necessary or to scribe and trim the abutting panel to fit ❷. To help create a tight joint, back bevel the stile and rails on the abutting unit before assembling the panel and before attempting to install on the wall. When the panel fits tightly at the corner, screw it securely to the wall ❸, ❹.

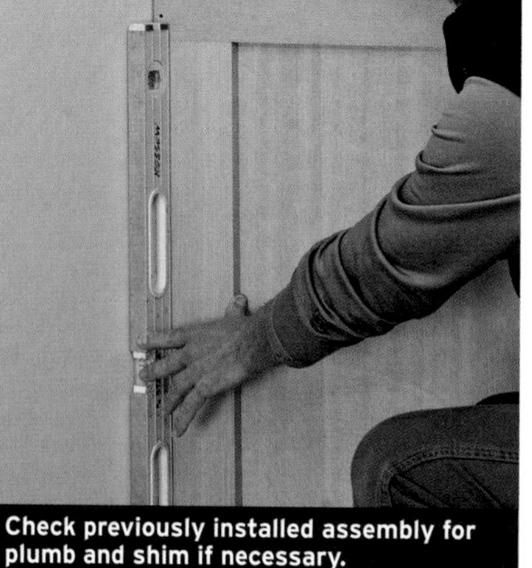

1 Check previously installed assembly for plumb and shim if necessary.

2 Scribe butting assembly to fit if necessary.

3 Drill angled counter bores into corner.

4 Drive screw into corner stud drawing joint tight.

WHAT CAN GO WRONG If the corner joint opens up during installation, you may be able to draw it tight without having to remove and retrim the second unit. Try drawing the joint up tight by backing off the screws and then driving in new screws at a sharper angle, pulling the second unit toward the first. The more you angle the screws, the more drawing power but be careful—a too-acute angle may cause the screw head to hit the sides of the countersunk hole making for an unsightly plug installation. If the gap is at the top, drive a shim between the wall and the previously installed panel to close up the gap.

DRAWING A PANEL TIGHT WITH AN ANGLED SCREW

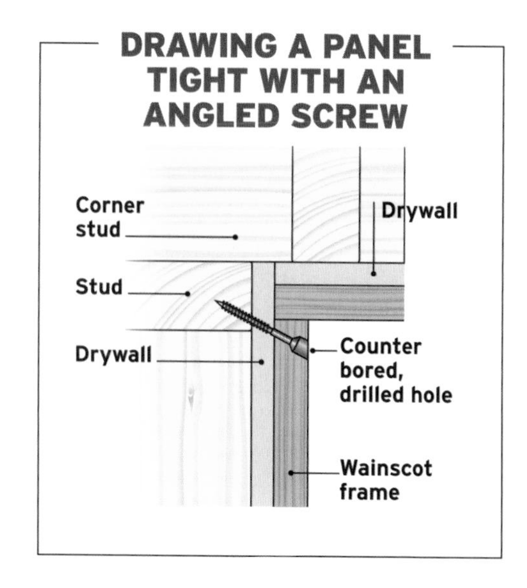

INSTALLING ASSEMBLED PANELS ON 90-DEGREE CORNERS

Begin by checking the outside wall corner for plumb. If it's out, add the amount of offset that occurs between the top and bottom of the wainscot to the total length of the assembled unit. Add the thickness of the unit plus another 1/8 in. for trimming and then mark the length on the stiles that will meet at the corner. Remove the stiles from the units and rip a 45-degree angle at the length mark. Re-assemble the mitered stiles to the units and then maneuver both sections of the corner into position and check the fit ❶. Moving one panel or the other in and out from the wall slightly will usually make the necessary corrections. Apply construction adhesive to the wall at the corner before final assembly to fill in any gaps between the panel and the wall. Finally, glue and nail the mitered corners together ❷.

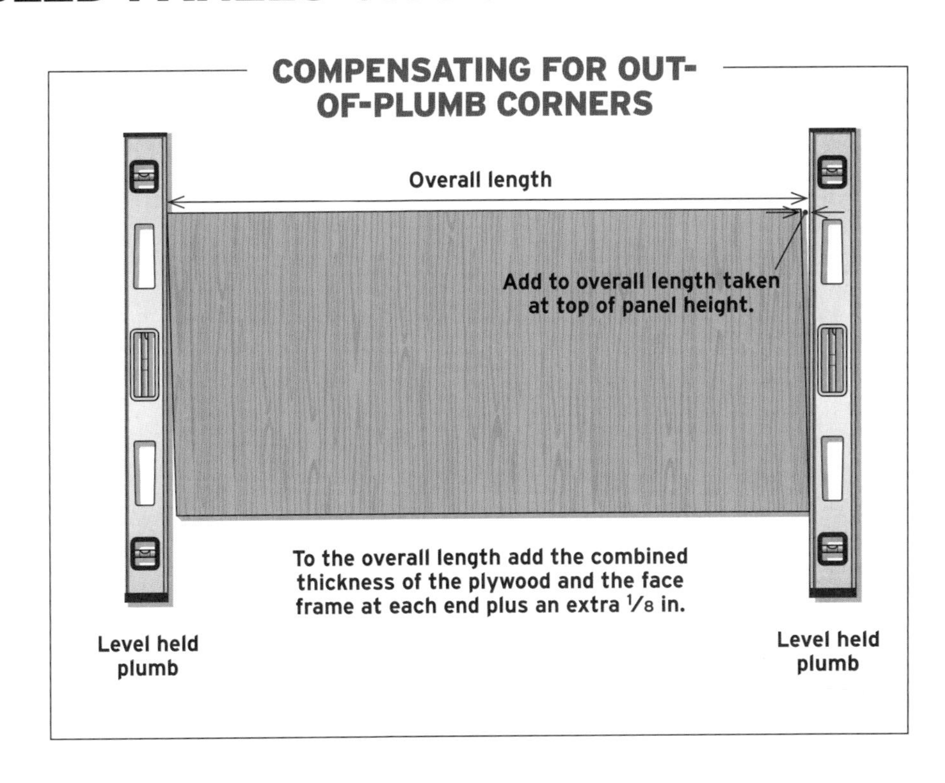

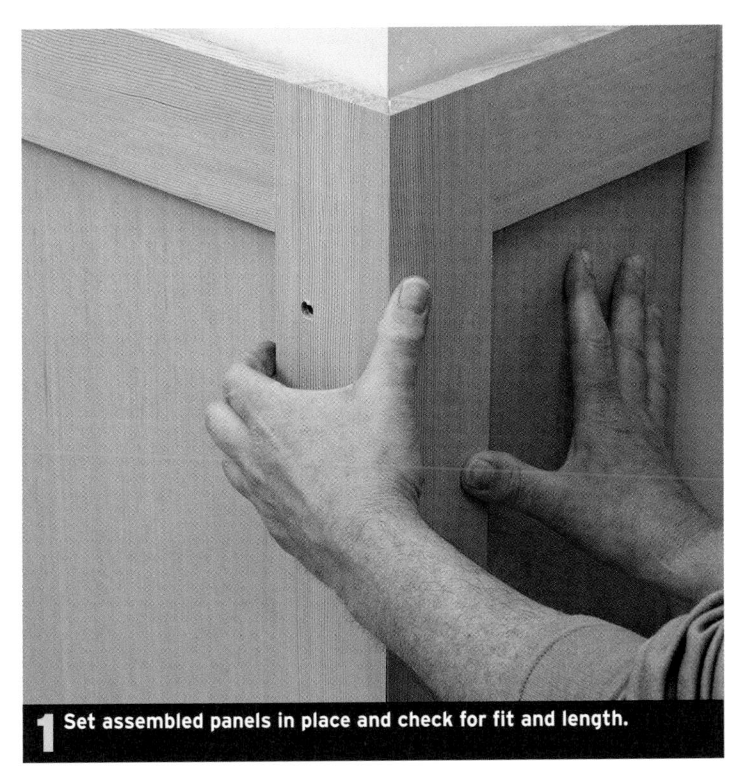
1 Set assembled panels in place and check for fit and length.

2 Nail the mitered corners together.

ASSEMBLED PANELS AROUND A BULLNOSE CORNER

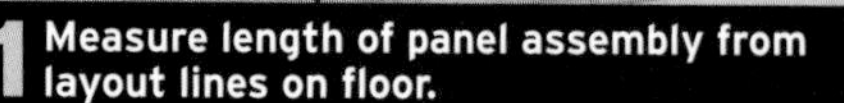

1 Measure length of panel assembly from layout lines on floor.

2 Set panel assemblies on top of layout lines [lines don't show in pix] and screw.

3 Mark width of corner transition piece onto story stick.

4 Transfer width of corner onto material.

5 Nail glued corner piece to adjoining pieces.

At a bullnosed corner a transitional corner piece with 22½-degree mitered edges takes the wainscot around the radius. To determine the overall length of the two wainscot units tack scrap pieces of plywood the same thickness as the wainscot back panel to the corner and to each side. Following the drawing at right, use a framing and a combination square to locate the corner points of the transitional piece. These points also determine the short point of the mitered stiles of the two wainscot units ❶. Cut the units to size and then set them in place to the layout lines and then screw the units to the wall ❷. Plumb them carefully so the width of the transitional piece will be the same along its length. Now mark the width of the corner piece across the corner on a story stick ❸ and cut it to width with a 22½-degree angle on each edge ❹. If you need to adjust the fit of the corner joints, either move the panel assemblies or recut or trim the angles on the edges of the transition piece to fit. When satisfied with the fit, glue and pin nail the miters together ❺.

DETERMINING CORNER POINTS AROUND A RADIUSED CORNER

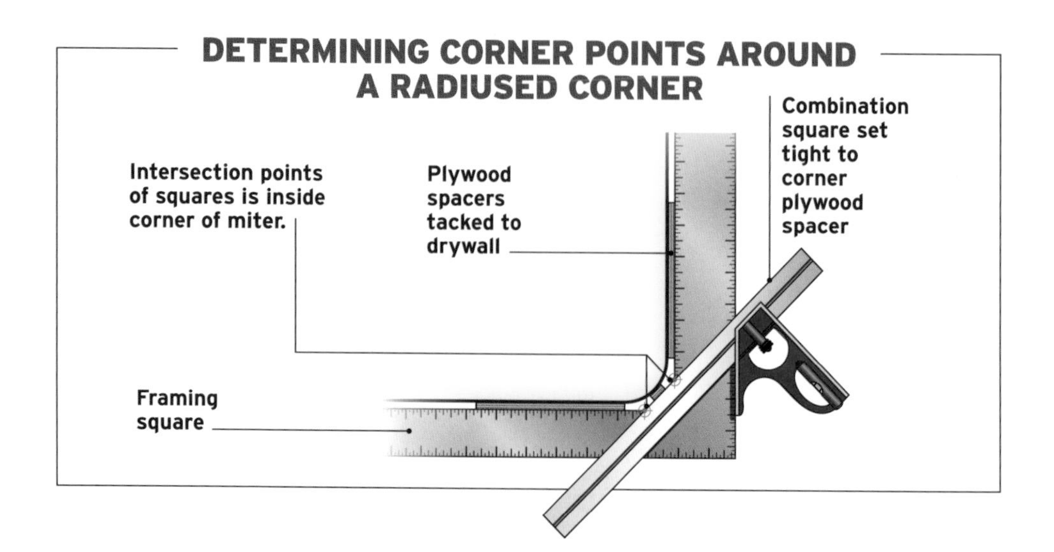

PRE-ASSEMBLED FRAME-OVER-PANEL

1 Apply glue to back of panel before applying to wall.

2 Nail in middle of a panel, hand pressing it flat to wall.

3 Angle nail into the drywall at panel edges and into wall framing.

4 Use story pole to mark out stile locations on rail.

In this option the paneling is applied to the wall and then a pre-assembled frame is applied over it. This method is quick, but if there are a lot of windows, standing moldings, heat registers, or other fixtures that must be fit, the piece-by-piece option may turn out to be the more simple and easy way to go.

The first step is to install plywood panels on the walls. Determine the layout of the panels in relationship to the positions of the stiles and to the wall studs and cut the plywood panels to size. Apply a squiggle of yellow glue to the back of the panel ❶ and nail it to the wall at the studs, aligning the vertical edge to the mark showing the centerline of a stile. If there are outlets, check to be sure the hole matches up with the box. Press the sheet against the wall to flatten the glue and then nail at an angle in the middle of the panels (into studs if possible) ❷. Next, nail along the entire perimeter of the panels. Since the edges likely won't be located over studs, angle the nails when shooting them in so they hold the plywood tight to the Sheetrock while the glue dries ❸.

Assembling the frames

Begin constructing the frame by marking the stile layout on the top and bottom rail, indicating the edges as well as the centerline of the slot ❹. Also mark the slot centerlines where the end of the rails intersects the end

>> >> >>

PRE-ASSEMBLED FRAME-OVER-PANEL (CONTINUED)

5 Mark centerline of slot on rail and the stile of corner assembly.

6 Align joiner on centerline and cut slot for biscuit in rail.

9 For a radiused corner, use a three-piece 22½ degree assembly.

10 Cut slot into corner assembly at centerline.

stiles ❺. Once the biscuit locations are marked, cut the slots ❻.

Assembling mitered stile corners

Even when building the frame on the wall piece by piece it is easier to pre-assemble the corners and install them before the rest of the frame. Mitered corner assemblies are easier to assemble accurately on the bench and they can be fastened to the wall using fewer fastenings. Create a 90-degree corner with 45-degree angles ripped on the edge of both pieces. Spread an adequate amount of glue on the mitered face to create a good bond ❼. Nail the glued miter ❽. The three-piece corner, designed to fit around a radiused corner, requires the edge rips to be cut at 22½ degrees ❾. Refer to the drawing on p. 198 to determine the size of the transitional piece on the three-piece corner—though you can omit the plywood panel backing as the panel will be already installed on the wall.

7 Spread glue on face of miter.

8 Nail together the miter.

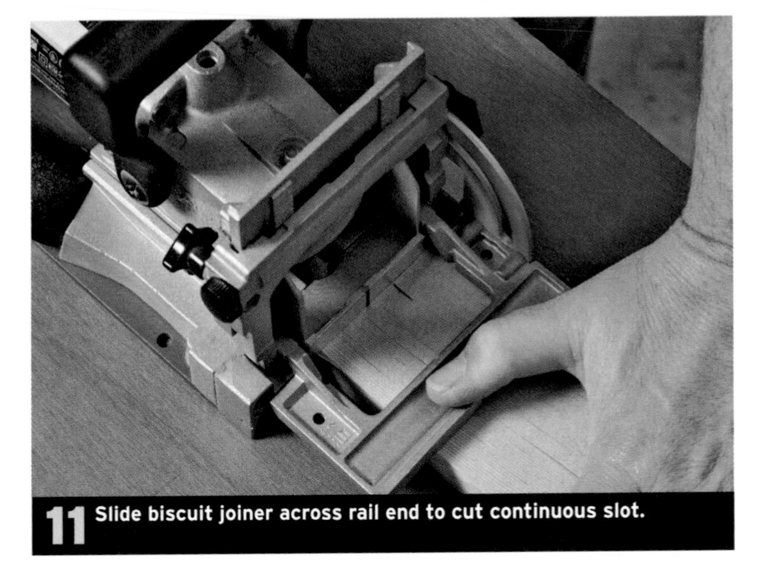
11 Slide biscuit joiner across rail end to cut continuous slot.

WHAT CAN GO WRONG

Note that on some biscuit joiners, to allow for a sliding motion you may have to remove (or recess) pins designed to prevent slipping.

If the corners are out of square you will need to adjust the miters accordingly before assembly by adding or subtracting a few degrees on the miter cuts.

Once you've marked the biscuit location of where the rail meets the mitered corner stile, cut the slot ⑩. In order to install the rails between two corner assemblies already attached to the wall, the rails must be able to slide down over the biscuits. To do this you must create a continuous slot along the rail ends. Use the biscuit joiner as a slotting router by plunging it in at the centerline as usual, but then making overlapping plunges to the outside edge of the rail ⑪.

INSTALLING FRAMES PIECE BY PIECE

1 Mark rail heights to paneling from story pole.

2 Plumb corner assembly while installing to wall.

3 Set bottom rail on leveling block and mark for length using a preacher.

4 Flatten one side of biscuit to allow for clearance in rail end.

5 Glue and insert biscuit into corner slot.

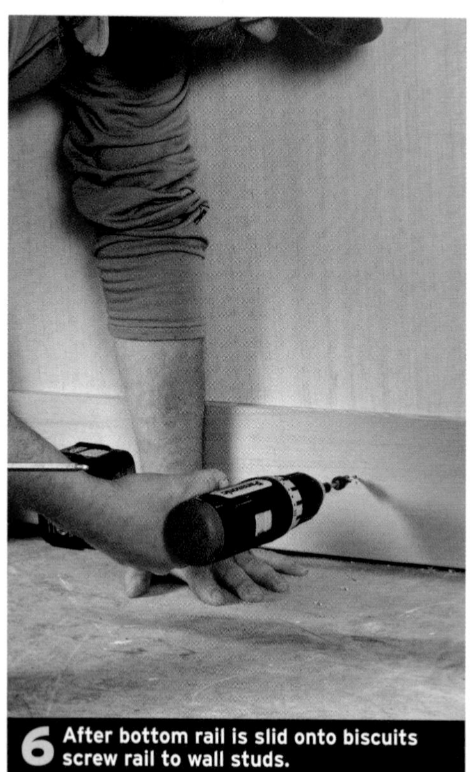

6 After bottom rail is slid onto biscuits screw rail to wall studs.

When the frame is applied to the wall piece by piece, all the stile and rail intersections are joined with biscuit joints. This keeps the faces of the frame flush and will not require as many fasteners to fasten the frame to the wall. It is easiest if the biscuits are located and their slots cut before installing the pieces on the wall. When using this method it is also easier to pre-assemble and attach the corners to the wall first. This makes for a very tight corner assembly and requires fewer fastenings to attach the corner pieces to the wall. The ends of the rails where they intersect the corner stile are slotted to enable them to slide over pre installed biscuits in the edges of the corner stiles.

After applying the plywood to the wall use a story pole to mark the top and bottom rail heights ❶. Attach the pre-assembled corner assemblies to the wall, plumbing plumb them in both directions with a level ❷. Set the bottom rail onto leveling blocks, butt one end to a corner stile and mark the other end for length with a preacher ❸. Cut the rail to length with a slight underbevel to enable the rail to slip easily between the installed stiles. Just before installing the bottom rail, glue a biscuit into the bottom slots on the corner stiles. Use white glue as its longer set up time allows you to adjust the pieces during assembly (it's more than strong enough for this application). To ensure a trouble-free fit, grind away a little of the biscuit to account for any unevenness in the long slot cut into the end of the rail ❹. Now apply glue to the rail end slots, insert the bottom rail above the protruding biscuits and then slide it down into position on the leveling blocks ❺. Then screw in to the wall studs ❻. To install the stiles glue biscuits into all the slots on the bottom rail, put a bit of glue into the stile's end slot and set each of the stiles in place pushing them tight over the biscuits ❼. Once the stiles are set in place, glue biscuits into their top slots and the top side edges of the corner stiles. Slide the top rail into position engaging all the biscuits ❽ and ❾. Screw the top rail to the wall studs making sure the joints stay tight. If necessary, drive a small pry bar into the wall studs immediately above the top rail to lever the joint tight while fastening to the wall.

WHAT CAN GO WRONG

This option will not work on irregular walls because the stiles are rarely located over a wall stud. This makes it difficult to draw the frame tight to the plywood panel where it dips into irregularities in the wall.

7 Install all the stiles before sliding top rail in place.

8 Slide the rail into position over the biscuits in the side of the end stile.

9 Slide the top rail into position engaging all the biscuits.

INSTALLING PRE-ASSEMBLED FRAME AND PRECUT PANEL

When inserting panels individually into a frame that you've already installed on the wall, you will be cutting the panels slightly smaller than the opening to ease the process. To cover the inevitable gap between the frame and the edge of the panel, you'll need to apply moldings around the perimeter of the panel. These moldings are generally rabbeted and therefore stick out past the face of the frame. A timesaving advantage of rabbeted moldings is that they can pre-assembled as a "picture frame" before installation. Be aware that perimeter panel moldings may not be compatible with non-ornamental trim-styles such as Shaker and Craftsman. Pre-assemble the frame with pocket screws and attach the unit to the wall with countersunk and plugged screws ❶.

➔ **See "Assembling the Frame-and-Panel Unit," p. 162.**

After the frame is secured to the wall measure and cut the individual panels 1⁄8 in. undersize in both directions so they will install without further trimming in between the frames. Attach the panels to the wall by applying yellow glue to the back of the panel and then pressing it in place against the wall

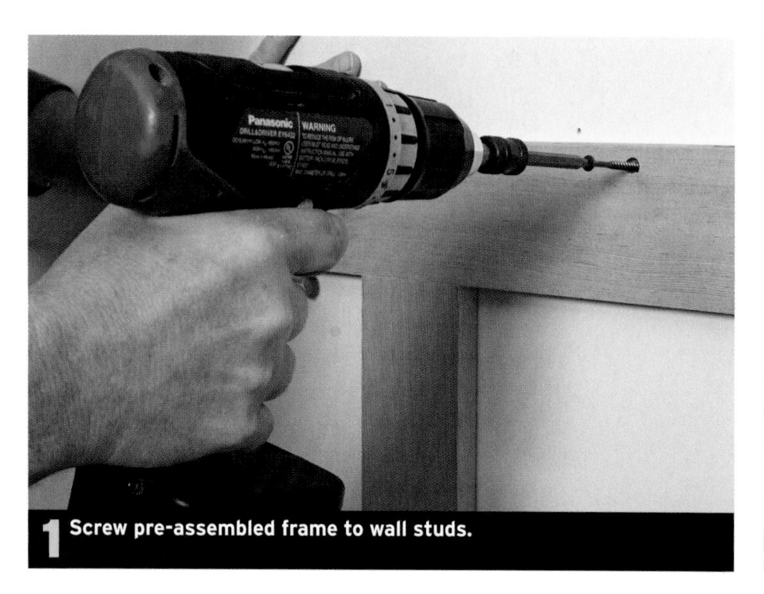

1 **Screw pre-assembled frame to wall studs.**

2 **Apply glue to back of panel and press in place.**

5 **Glue and pin nail miters.**

6 **Clamp corners until dry.**

❷. Nail around the edges of the panel close enough to the edges so that the perimeter moldings will cover the nail holes ❸.

Making pre-assembled perimeter moldings

Rather than installing the perimeter moldings piece by piece on the wall framing, you can save time by pre-assembling the moldings into a "picture frame" and then installing them as a unit. To determine the measurement between the inside of the rabbets on the moldings, measure the length and height of the panel opening and subtract 1/8 in. This will allow 1/16-in. play on all sides. Be sure to cut the opposing pieces to exactly the same size to ensure that the unit comes out square and has tight joints ❹. Glue and nail the miters together ❺, then clamp and let the glue set up before installing the unit to the wall frame ❻. To ensure that the assembled moldings line up along the length of the frame and panel unit, keep the bottom rebates of the moldings tight to the bottom rail. Center the frame in the opening and pin nail it in place ❼.

3 Nail edges of panel at an angle into the Sheetrock.

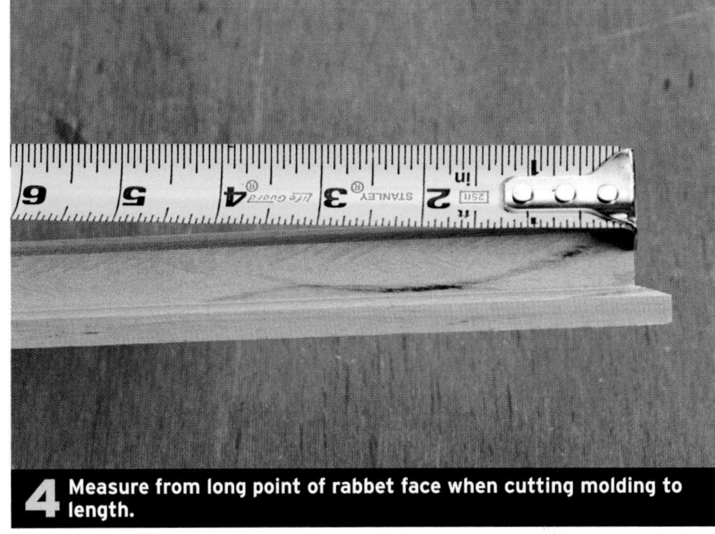

4 Measure from long point of rabbet face when cutting molding to length.

7 Nail pre-assembled molding unit to the frame.

CEILING ELEMENTS

OF ALL THE FINISH WORK WE MAY install in a house, ceiling coverings may have the greatest impact on the feel and style of a room. A trimmed ceiling's woody colors, rich textures, and pleasing geometrical patterns dramatically change the feeling of a room—especially in comparison to the unadorned, blank canvas of Sheetrock. In a renovation project, these trim treatments are also an effective way to cover an old or damaged ceiling.

Make no mistake, ceiling treatments may be one of the more challenging and tedious trim projects you might undertake. This isn't due to unusual complexity or unique challenges, but simply because of the physical demands of working for long periods over one's head. While it might not drive you crazy as it did Michelangelo, you will definitely experience a sore neck and arms by day's end. However, if you rig scaffolding and use efficient measuring and installation techniques as presented in this section—and especially if you find another person to help you at certain points—the work will be decidedly more bearable.

TONGUE-AND-GROOVE

PANELED CEILING

COFFERED CEILING

EXPOSED BEAMS

POSTS

Photo by David Duncan Livingston

TONGUE-AND-GROOVE CEILINGS

Photo by Tim Street-Porter

This room is anything but traditional, but the warm woodtones contrasted with the shiny, modern stainless steel creates a strong statement.

Traditional bead board is tongue and groove with a bead at the joint. This beadboard is actually a scored plywood panel that's easy to install and hides a multitude of sins.

Photo by Kevin Ireton, courtesy *Fine Homebuilding*

A run of t&g (tongue-and-groove) boards—relatively economical and straight-forward to install—instantly makes a room feel cozier by adding warm color tones and a soothing-to-the-eye, regular arrangement of joint lines to the ceiling. Unlike the flat, white surface typical of painted Sheetrock, tongue-and-groove boards break up the natural light brought in through the windows—reducing glare and reducing the reflection of harsh white light throughout the room. Though a raked ceiling—and especially one with a change in rake (also called a pitch break)—does add some challenges to the fitting and installation process, a tongue-and-groove ceiling is still one of the easiest ways to transform a room completely into a cozy, appealing space.

Photo by Charles Miller, courtesy *Fine Homebuilding*

In this kitchen tongue-and-groove combines with exposed beams to give this kitchen both a traditional look and a cozy feel. Painting both the same color integrates the two elements.

FITTING STARTING BOARDS ON A FLAT CEILING

1 Hold board tight to wall and scribe wall outline onto it.

2 Use an inverted saber saw to backcut the scribed line.

3 Clean up the edge with a block plane.

4 Push the board tight to the wall and nail to the ceiling.

Installing tongue-and-groove boards on a ceiling is relatively straightforward. After setting the first board in place (with the tongue looking at you), the process is simply a matter of cutting the next board to length and fitting it into the last installed piece. You'll continue like this, making sure to nail the boards into the ceiling framing (which is generally underlying Sheetrock), until you reach the other wall (or peak of the ceiling).

Begin by selecting a straightedged starting board. Cut it to length, fit the ends, and hold it tight to the wall (be careful not to force it to bend to any curvature of the wall) and look along the edge for gaps. If the material is tongue and groove, orient the groove side toward the wall. If any gaps appear due to any irregularities or bows in the wall, use a scribe to draw the outline of the wall onto the board ❶. Cut to the scribed line with an inverted saber saw held at a slight undercut angle ❷.

Then use a block plane to trim fit the edge ❸. Install the starting board tight to the wall, nailing it to the ceiling framing ❹. This process ensures that the following ceiling board courses will have a straight edge to start off from.

FITTING STARTING BOARDS ON A RAKED CEILING

1 **Use a sliding bevel to determine ceiling intersection angle.**

2 **Transfer bevel setting directly to saw blade.**

3 **Rip the board face side up to reduce splintering.**

4 **Hold the board in place and check for fit to wall.**

Begin by determining the intersecting angle of the wall and ceiling with a sliding bevel ❶. Transfer this angle directly to the saw ❷ or, alternatively, determine the angle in degrees set the saw's bevel indicator to that measurement. To ensure a tight fit, set your tablesaw several degrees more than the angle recorded by the bevel gauge. Now rip the board to the required angle orienting the board face up to reduce splintering of the edge ❸. Cut the board to length and fit the ends. Temporarily tack or hold the board in place and check the fit of the edge ❹. Scribe and trim the edge as necessary.

Be careful when handling the cut board, as the trimmed edge can be very sharp. Finally, nail the board in place through the face and tongue into the ceiling framing ❺.

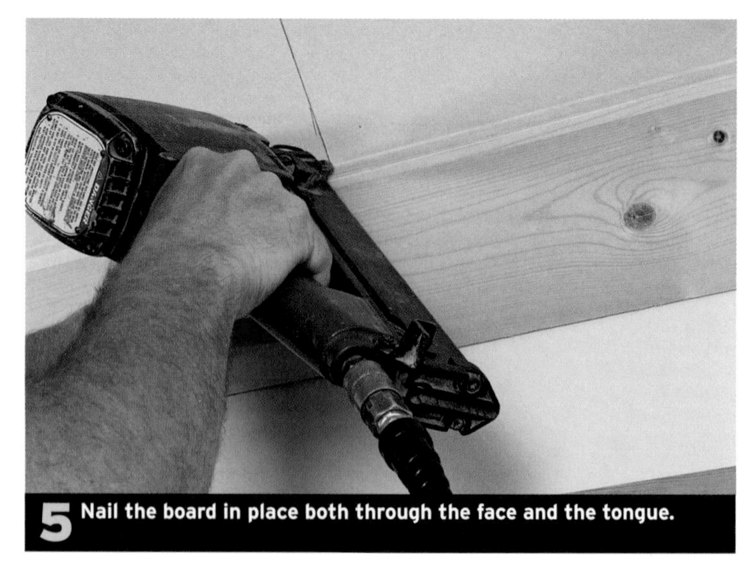

5 **Nail the board in place both through the face and the tongue.**

PULLING TOGETHER STUBBORN TONGUE-AND-GROOVE

1 Use a piece of scrap to protect the tongue on ceiling board.

2 Drive a pry bar into ceiling framing and lever joint closed.

3 Screw block to ceiling framing and drive shims to tighten joint.

There are times when the tongue-and-groove boards do not want to close up. This is most often due to the boards being either bowed or warped—a problem that can be dealt with. Sometimes, however, the boards are incorrectly milled and little can be done other than to return them. Here are three different methods for tightening the joints: Place a piece of scrap against the board and then tap the scrap's edge with a hammer to snug the groove into the tongue ❶. If the joint does not stay tight, drive a chisel or pry bar into the ceiling framing next to the scrap and lever the piece into place, nailing it while you hold the joint tight ❷. For stubborn pieces, screw a block into the ceiling framing and insert wedges between it and the ceiling board. Pounding the wedges will generally force the board into compliance long enough to nail it in place ❸.

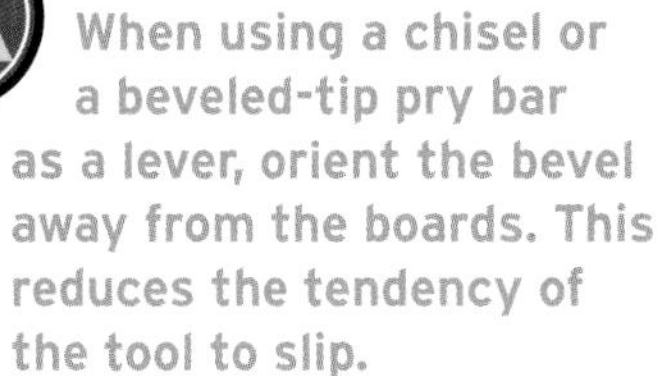

TRADE SECRET

When using a chisel or a beveled-tip pry bar as a lever, orient the bevel away from the boards. This reduces the tendency of the tool to slip.

CUTTING OUT FOR A CEILING FIXTURE

1 Measure distance to box from wall.

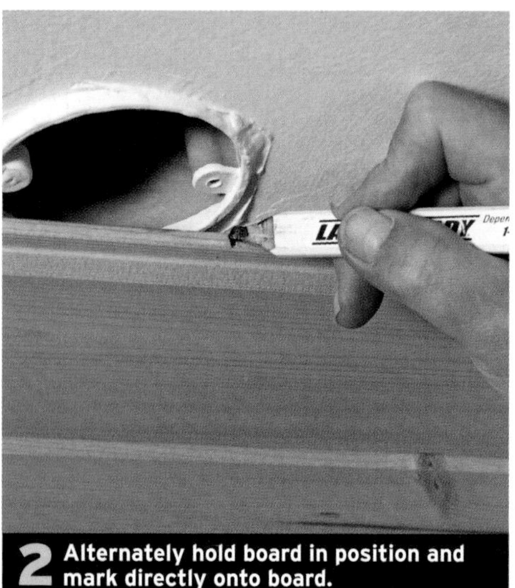
2 Alternately hold board in position and mark directly onto board.

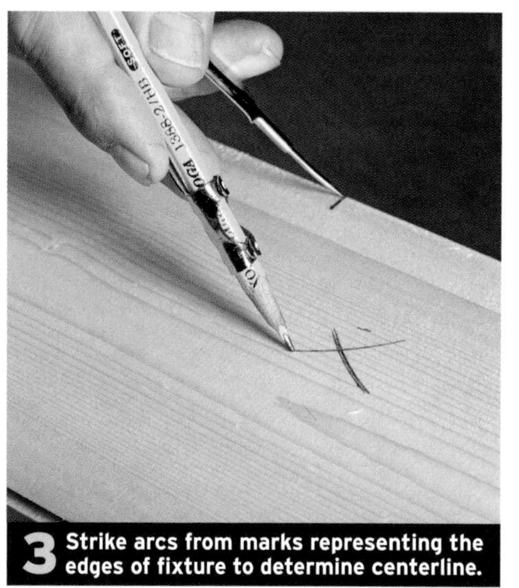
3 Strike arcs from marks representing the edges of fixture to determine centerline.

4 Draw centerline adding additional piece if center of box is off board.

WHAT CAN GO WRONG

If the boards are tongue-and-groove, make sure when measuring to the adjacent board to measure to the shoulder of the tongue. Then the dimension can be pulled directly from the edge of the grooved side of the board without having to subtract the width of the tongue.

Recessed light cans and rectangular electrical boxes are the most common ceiling fixtures that board ceilings have to be cut around. Check to make sure the fixtures protrude down from the ceiling enough to allow the trim to cover up the hole after the ceiling boards have been added. If not, either the fixture will need to be lowered or an extension added. To determine the box location either measure from a wall or adjacent board end ❶ or hold the board in place and mark the edges of the box directly on the board ❷.

Strike an arc from the two edge points to find the centerline of the circular box ❸. Now measure the distance from the edge of the box to the shoulder of the tongue on the board immediately adjacent to it. If the center point of the box will land on the next board over, position a piece of scrap next to the box location and use a square to draw the centerline on both boards ❹, ❺. Transfer this distance to the centerline on the board and mark it ❻. Now set a compass to the radius of the box, put the point on the mark representing the edge of the fixture and strike a point on the centerline to represent the center point of the fixture ❼. Reverse the compass and draw in the circular outline of the box ❽. Use a saber saw to cut out the waste ❾. After nailing the first piece in position, repeat the process for the next board and nail in place ❿.

If the fixture ends up centered on a board, it may take up the whole board and require cutting the board into two pieces to install.

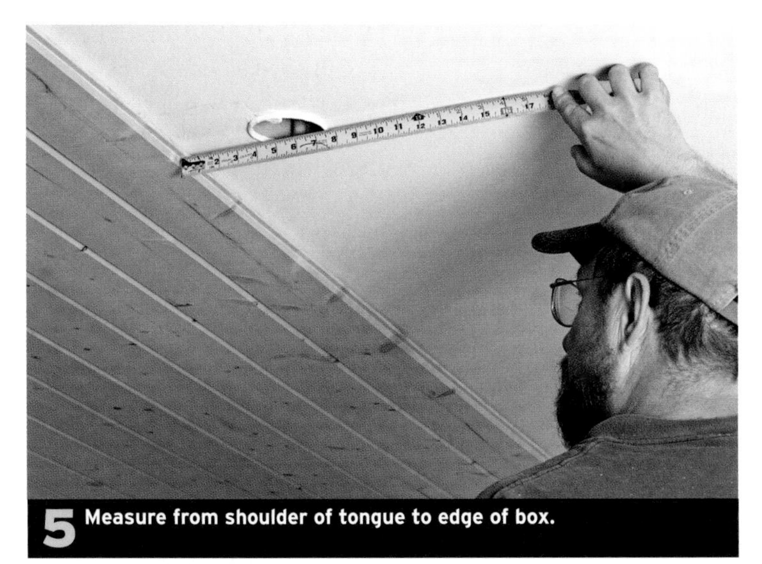

5 Measure from shoulder of tongue to edge of box.

6 Mark edge of box onto centerline.

7 Set compass to box radius, mark box center off the edge mark.

8 Reverse compass and draw in the outline of the box cutout.

9 Cut out waste with a saber saw.

10 Repeat the process for the second piece and nail in place.

FITTING AN ANGLED CEILING BREAK

1 **Determine the miter angles with a proportional divider.**

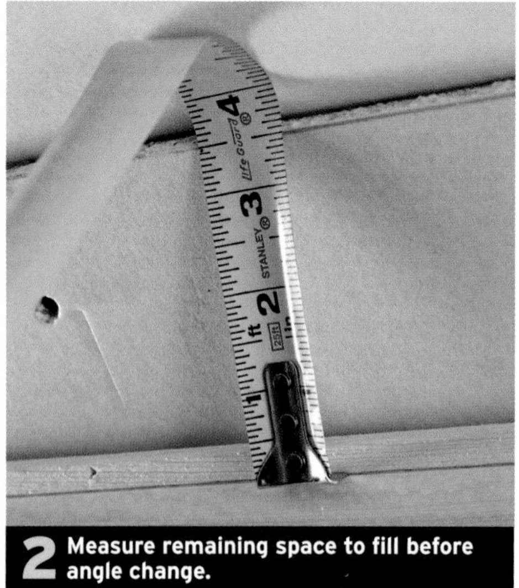

2 **Measure remaining space to fill before angle change.**

3 **Use scrap pieces to check for correct size and miter angle.**

A ceiling that has two different pitch angles that meet—and show—requires a different strategy than a flat ceiling. This installation problem can be dealt with in two ways. The simplest is to nail a small trim strip over the joint. While this may seem like the easy way out, it can also be the best way to deal with a pitch break as it not only is quick to do, but it conceals any potential gaps that might arise from the shrinkage or expansion of the boards. Usually the smaller the trim piece the better it will look.

The other alternative, of course, is to bevel the boards and fit them together. Done well, and assuming the joint stays gap free, this strategy draws less attention to the break in pitch, blending the joint seamlessly into the rest of the ceiling surface.

Begin the installation of the boards covering the pitch break by using either a bevel gauge or a proportional divider to determine the angle of the break ❶.

Measure the distance to the joint intersection from the last board installed at various points along its length. Use the smallest dimension obtained for the width of the first board ❷. If the space is tapered from one end to the other, adjust the spacing of the preceding boards slightly to make up the difference to avoid a difficult tapered rip.

➔ See "Installing a Closing Board to an Irregular Wall," p. 153.

Rip two pieces of scrap to the predetermined angle and try them on the ceiling before ripping the full-length boards ❸. After cutting the first mitered break board to the correct length, rip it to width at the predetermined angle and nail the first board to the ceiling ❹. Next, rip the matching angle on the edge of the adjoining board and hold it in place to check for fit. The joint can be tightened while nailing by using a small pry bar driven through the Sheetrock into the ceiling framing to pry the board over ❺.

Uneven ceiling

If the ceiling is uneven, it may be necessary to insert shims between the boards and ceiling to bring the mitered edges more in line with one another ❻.

Using double screws

The miter joint along a pitch change can be tightened by using double threaded screws. Do not nail the boards next to the miter but insert double threaded screws ❼. When the screw is extracted, the bottom thread starts to draw the screw from the framing. The upper portion, which has an increased diameter and a reversed screw pitch, moves the attached board further away per revolution than the screw itself is being extracted and holds the board in that position ❽.

DOUBLE THREADED SCREW

Larger diameter, bigger screw pitch and opposite lower portion

Smaller diameter than above

4 Nail the first break board to the ceiling framing.

5 Use a pry bar to tighten joint while nailing second break board to framing.

UNEVEN CEILING

6 Shims are used to even out uneven ceiling.

7 Screws ready to drive into boards next to open miter.

8 Miter joint after screws have been used to close miter joint.

A small molding can be applied over the joint. **Not only does this provide an additional visual element it hides any gap caused by seasonal wood movement.**

INSTALLING A CLOSING BOARD

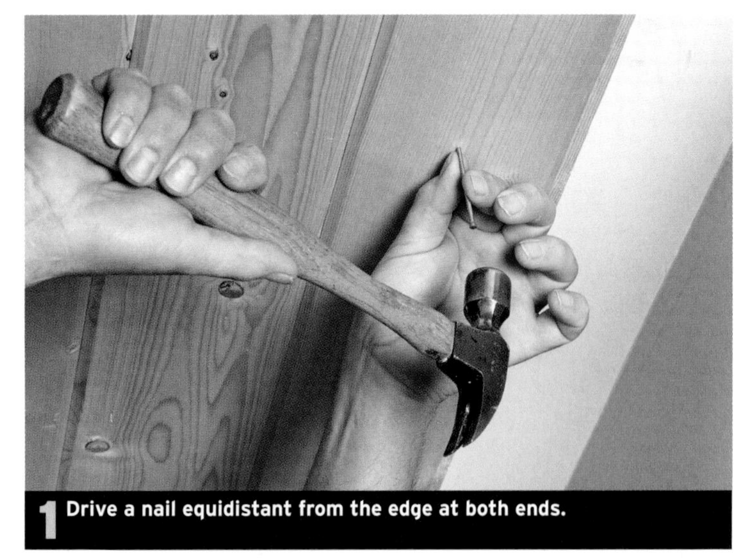

1 Drive a nail equidistant from the edge at both ends.

2 Set compass to width of opening.

5 Reset the compass to the width between the mark and the wall.

6 Using this width scribe the wall to the board.

If the wall meeting the closing board is straight and runs parallel to the board, simply rip the board to width and remove the back half of the groove on the tablesaw. (This allows the board to be nailed in place without engaging the tongue.) When the wall is straight but tapers from one end to another relative to the closing board, measure each end of the opening and cut the taper on the board with a circle saw. Alternately, the taper can be eliminated by adjusting the spacing of the previous boards.

When the closing board needs to be scribed to fit an irregular wall, the process becomes a bit more involved. Begin by marking a point equidistant from the edge near each end of the last installed board. Drive a nail into these points ❶. The nails will align the boards parallel to one another during the scribing process. Now set a compass to the width between the shoulder of the tongue and the wall ❷. Cut the closure board to length and butt it's edge to the nails, temporarily nailing it in place or using braces extended from the floor to hold it in position ❸. Using the compass to transfer the previously determined width of the closure opening to the board ❹. Now, leaving the board in place, reset the compass to the width from the just struck mark on the board to the wall ❺. Using this width scribe the entire length of the board ❻. Now remove the board and cut to the scribed line using a saber saw held at a slight angle to create an undercut ❼. After cutting attach the board to the ceiling by nailing through the face into the ceiling framing ❽.

3 Butt closure board to nails and temporarily hold in position.

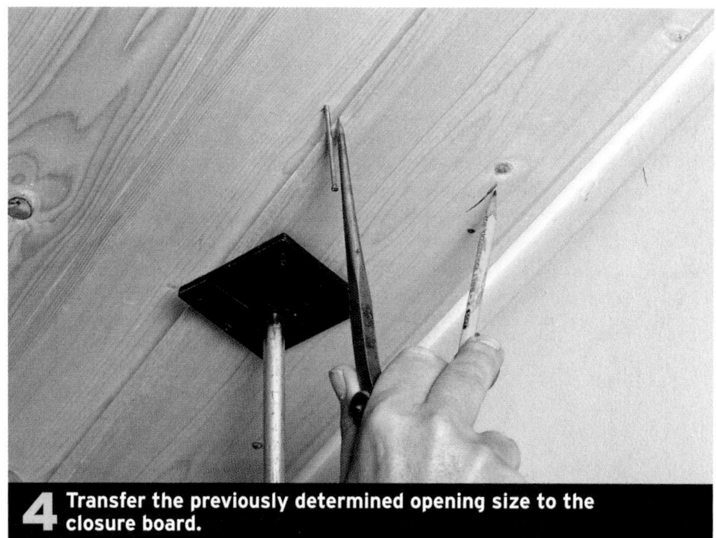

4 Transfer the previously determined opening size to the closure board.

7 Use inverted saber saw to back cut to scribed line.

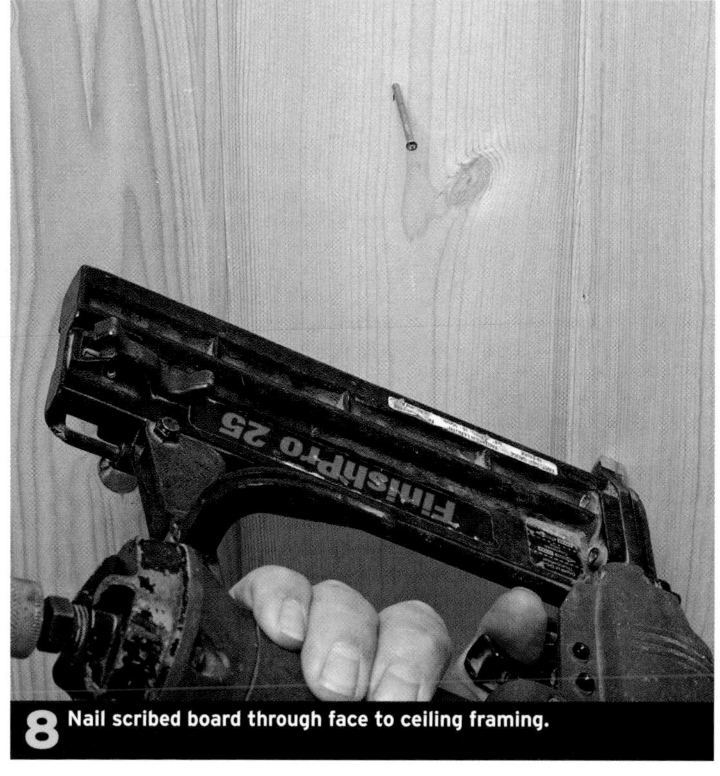

8 Nail scribed board through face to ceiling framing.

WHAT CAN GO WRONG It often happens that the ceiling boards will not run parallel with the wall coming up. To eliminate having to cut a tapered closure board, start measuring the distance to the wall from the last installed board at least 4 ft. before coming to the wall. By not driving the joints tight at one end or the other, the difference can slowly be eliminated by the time the last board is reached.

LOCATING CEILING JOISTS AND LAYOUT

Begin the layout by locating and marking out the ceiling framing throughout the room. Transfer these measurements to a scale drawing of the room on grid paper and draw in framing. Now work at laying out the trim boards to fit the size and shape of the room while creating a pleasing arrangement of inset panels—at the same time attempting to land the long edges of the sheets on the ceiling framing as much as possible. If there is no choice but to land a long edge between the frames, plan to use construction adhesive between the Sheetrock and the plywood to secure the panel.

Once the layout of the ceiling grid has been determined, use a chalk line to mark reference lines for orienting the plywood to the ceiling across the length and the width of the room ❶. Taking the time to install the plywood accurately to these lines will ensure that the seams run straight and can be used as a centerline off of which the beam sides can be marked onto the ceiling. At this time also determine the ceiling joist locations and record these locations either on a piece of paper or stick pieces of blue tape to the wall at the joist locations ❷. Because you'll use construction adhesive between the plywood and the ceiling it's not critical that all the nails hit on joists—though the more you can hit after the sheets have been tacked in place, the better.

SAMPLE LAYOUT OF PANELED CEILING WITH TRIM

This diagram shows the relationship of the joists to the trim used to create the coffered ceiling on the following pages. It is important to make a diagram of your own project and carefully lay out the lines on the ceiling before beginning installation.

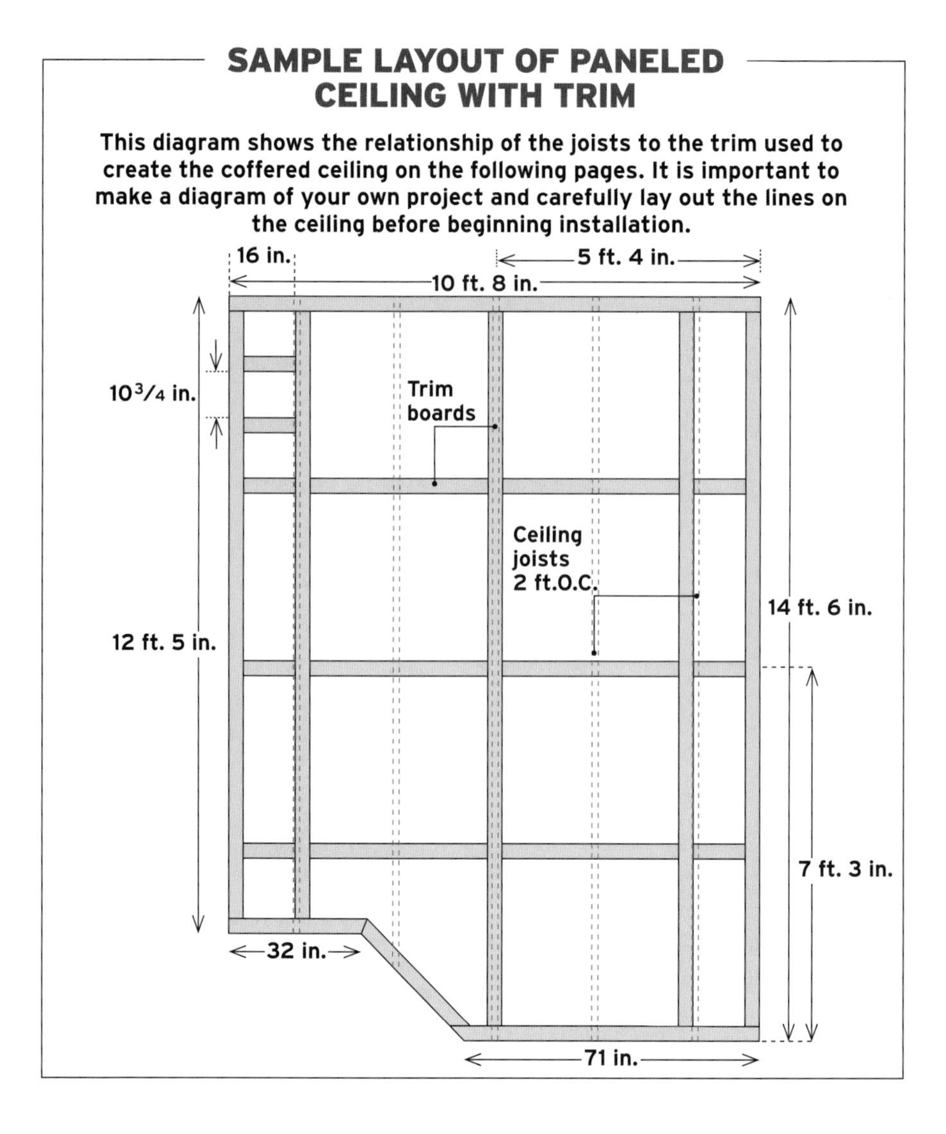

1 Chalk lines where edges of plywood will lie.

2 Use stud finder to locate joist locations.

CUTTING OUT FOR A FIXTURE IN A PANELED CEILING

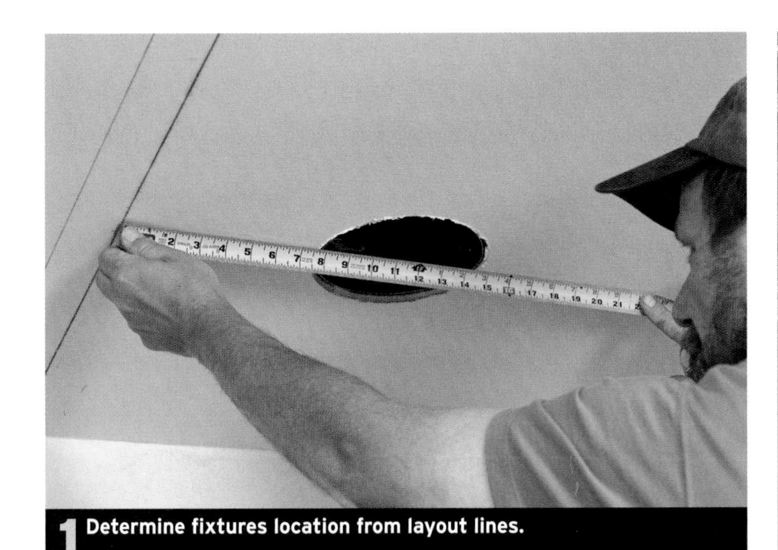

1 Determine fixtures location from layout lines.

2 Transfer measurements to back side of plywood.

Lay out any ceiling fixtures by measuring from the plywood location lines on the ceiling ❶. (You can also measure from an already installed sheet). Transfer these measurements to the back of the sheet ❷. When in doubt of the center point on a round fixture, measure to the outside edges of the fixture and draw in a box using those points. Then draw in the circle using diagonals across the corners of the box to determine the center point ❸.

After the fixture location has been double-checked, drill a starter hole in the sheet for the saber saw blade and cut the hole out with a saber saw ❹. If the hole is the correct diameter but is slightly out of alignment with the fixture, the location of the hole may be altered by trimming the edges of the sheet that will get covered by the trim boards and sliding the sheet over.

3 Draw in outline of fixture.

4 Cut fixture hole in plywood.

INSTALLING PLYWOOD PANELS

1 Attach site built support to ceiling to support end of plywood sheet.

2 Apply construction adhesive along inside edges of layout and on approximate joist locations.

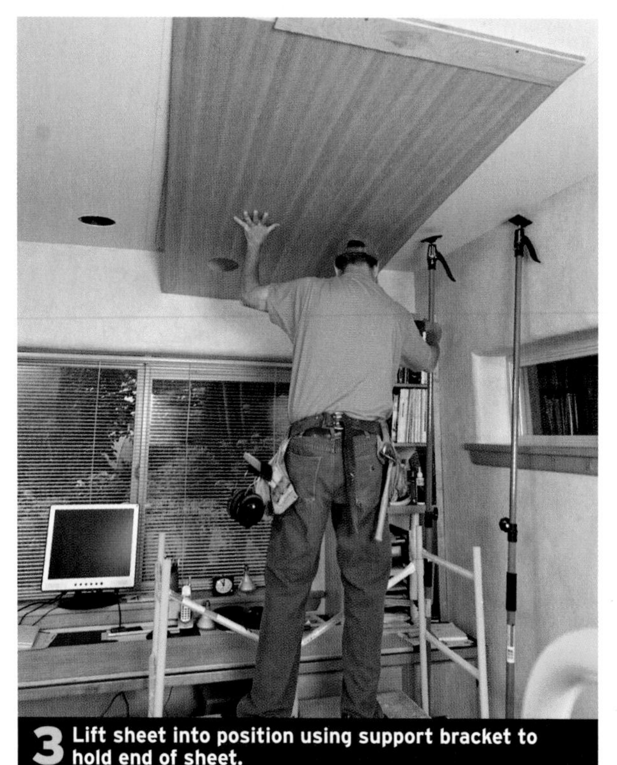

3 Lift sheet into position using support bracket to hold end of sheet.

Start by setting up scaffolding so you can easily reach the ceiling with your arms extended above your head. If you are going to have to work alone, make up a support bracket from a scrap of plywood and a 1 in. by 3 in. board and temporarily attach to the ceiling framing to hold up one end of the plywood for you ❶. (See "Trade Secret" on the facing page.)

Apply a bead of construction adhesive inside the chalked lines on the ceiling representing where the edges of the first sheet will lie and along the approximate locations of the ceiling joists ❷.

When applying the adhesive, keep the bead 1/4 in. thick or less to ensure that it will spread out and not create a bump in the plywood or prevent it from lying flat and tight to the ceiling. Now lift the sheet into place and slip one end into the support bracket ❸. Using either commercially made adjustable support poles or site-built 2X4 support fixtures to hold the other end of the plywood tight to the ceiling, align the plywood sheet to the chalk layout lines ❹.

Start at one end of the sheet and nail around the perimeter of the entire sheet every several inches ❺. If the edge of the sheet is not located over a joist nail through the plywood and into the Sheetrock at an angle to increase the holding power of the nail. Measure over to the ceiling joists in the center of the sheet and shoot nails into them every several feet while you push the panel tight to the ceiling. Be sure to press the glued areas of the plywood sheet tight to the ceiling to ensure that the adhesive makes contact and the sheet is tight to the ceiling ❻. Repeat these steps on the following sheets ❼.

SITE-BUILT SUPPORT STRUCTURE

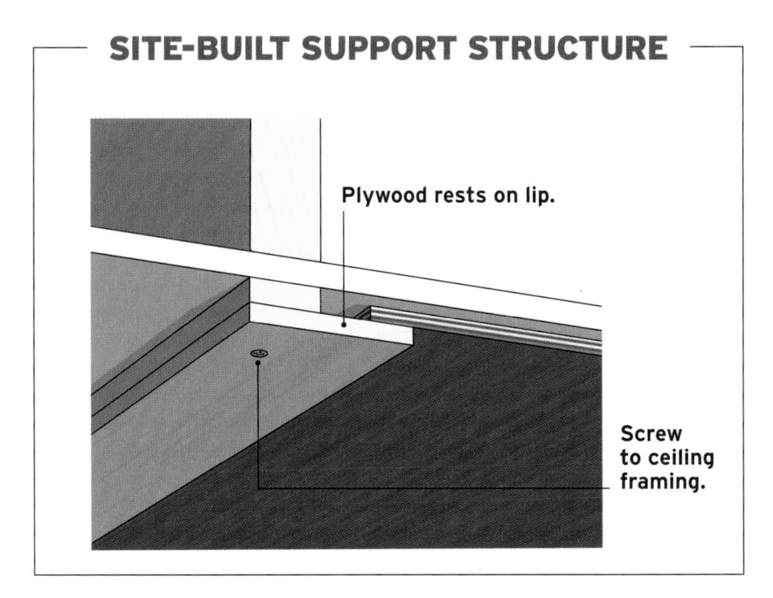

TRADE SECRET When doing a ceiling solo, use a site-built support fixture as shown in the drawing at left. Temporarily screw the fixture to the ceiling joists to hold the other end of the plywood in position while you fasten the plywood sheet to the ceiling. When installing sheets of plywood in a row, fasten the fixture on top of the previous sheet keeping it close enough to the end of the sheet to allow the screw holes to be covered by the applied trim board.

4 Use supports to hold sheet while aligning to layout lines.

5 Nail around the perimeter of the sheet every several inches.

6 Measure over to locate joists and nail every 2 ft.

7 Repeat steps on following sheets.

INSTALLING TRIM BOARDS ON A CEILING

You can create a visually effective coffered ceiling with only a modest outlay of materials and labor by utilizing sheets of plywood for the inset "panels" (the sheets really cover the entire ceiling) and defining the grid work of the coffer by applying flat stock. Moldings, either inset or rabbeted, add a surprising amount of texture and visual interest to the treatment. The trim boards must not only cover all the joints between the plywood panels but they must also create a geometrically pleasing network of inset panels.

Begin by laying out the locations of the trim boards that cross the ceiling on the installed ceiling panels.

➔ See "Locating Ceiling Joists and Layout," p. 188.

Because the seams of the installed plywood also represent the trim board centerlines, mark out both sides of the trim boards with a centering batten ❶. Double-check to make sure the trim lines run parallel to one another and join at right angles ❷.

Now install the trim boards that run around the perimeter of the ceiling. You can install them all at once or only as many as necessary to enable a section of the ceiling to be completed before moving the scaffolding. If the wall is irregular, rip the perimeter boards 1/2 in. oversize to allow additional material for scribing and cutting to fit. To scribe the cut line, measure from a parallel layout line or an already installed ceiling trim board to indicate where to secure the perimeter board temporarily to the ceiling next to the wall ❸. To ensure that the panels end up the same size make sure that the perimeter board ends up parallel to the other board. This is done by measuring over from the next board or an adjacent reference line. Mark the perimeter board to the desired width, set the pencil on the scriber to this mark and the point against the wall, and then scribe the board along its length ❹. Once the perimeter boards have been cut to fit, apply a thin coat of construction adhesive on the backside near each end of the board before you attach them to the ceiling.

Keep the nails close to the edge of the board where they will be covered by the applied rabbeted molding and set them at an angle into the ceiling ❺. If a splice is required to lengthen a board, cut the joint at 30 degrees, apply yellow glue to the splice and nail through the splice into the plywood ❻. Once the boards are in place, install the trim boards that will run the full length of the room in a similar fashion. Then install the cross boards. The order of installation for your particular layout will become obvious as the work proceeds. Cut the intersecting boards a little longer than necessary and then mark them to length in place ❼. A slight undercut at each end of the board will allow the butt joints to be started and then compressed as the board is pushed up into place ❽. After all the boards are installed, you can move on to installing the pre-assembled moldings.

➔ See "Installing Pre-Assembled Panel Moldings," p. 196.

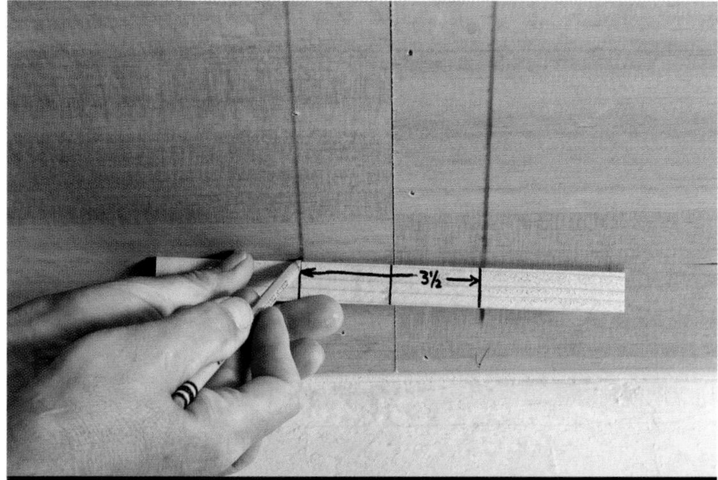

1 Use a centering story stick to layout edges of board from centerline.

4 Temporarily hold board in position while scribing to wall.

2 Mark board layout onto ceiling.

3 Measure from layout line or installed board to position perimeter board temporarily.

5 Push board tight to ceiling and nail at angle into ceiling plywood.

6 Nail spliced joints at angle through miter.

7 Mark individual boards to length.

8 Push board up into place for tight fitting joints.

INSTALLING CEILING PANEL TRIM MOLDINGS

On a complex ceiling panel, it may be easier to install the panel moldings piece-by-piece rather than as preassembled units. This may require more climbing up and down from the scaffolding or ladder but will certainly be less work than having to disassemble a finished molding that doesn't fit.

Start by measuring the length required for the first piece and add several inches to allow for marking in place ❶. Cut the appropriate angle (here it's 45 degrees) on one end, hold the molding in place with the angle cut set in the corner, and mark the other end for length ❷. Continue by cutting a matching angle on the next piece of the inside corner, butting it into the corner joint and then marking the other end to length ❸. To cut an outside corner, first determine the angle with a bevel indicator ❹.

In order to mark both halves of the outside corner to length prior to installing the first half of the outside corner, mark the second half to length ❺. Once the pieces fit, glue the joint and nail in place ❻. Note that the closure piece will be mitered on both ends ❼.

1 Measure and add several inches to the length of the first piece.

4 Determine corner angles by using a miter indicator.

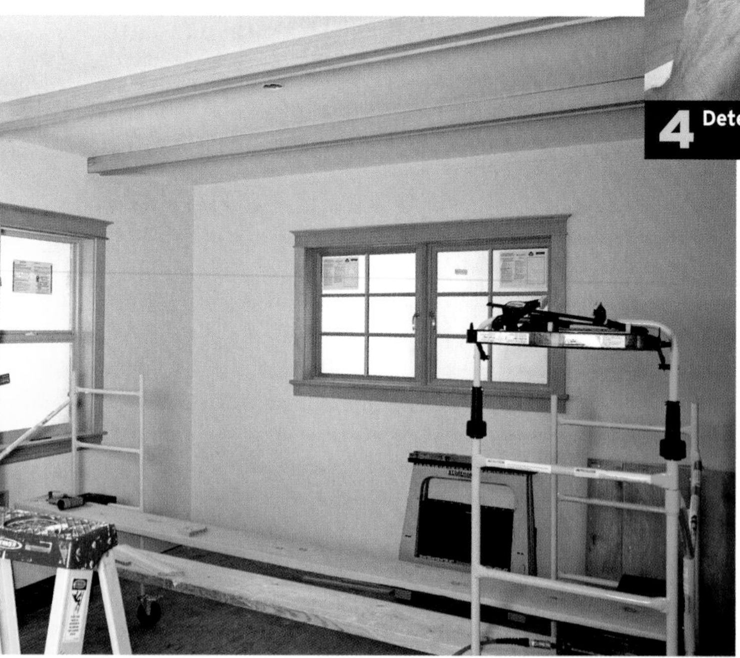

Scaffolding Installing ceiling elements requires constant moving back and forth to fit the boards together. Doing this on a ladder, especially when you are working solo, is a thankless job. It's best to have a second person to help you, but even with help, however, working over your head comes down to hard work. For ceilings over 8 ft. high, the job will go easier and

2 Butt the mitered end into corner and mark other end to length.

3 After nailing first piece in place repeat process and install second piece.

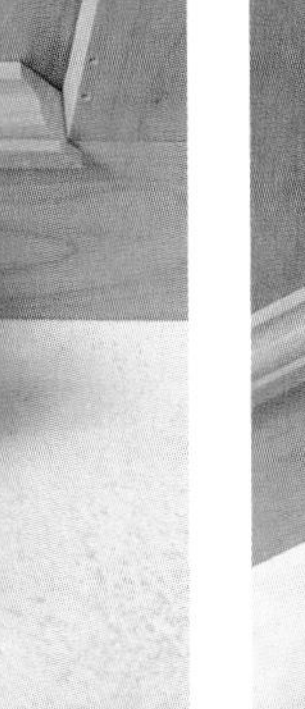

5 Before installing first half of outside corner mark second half to length.

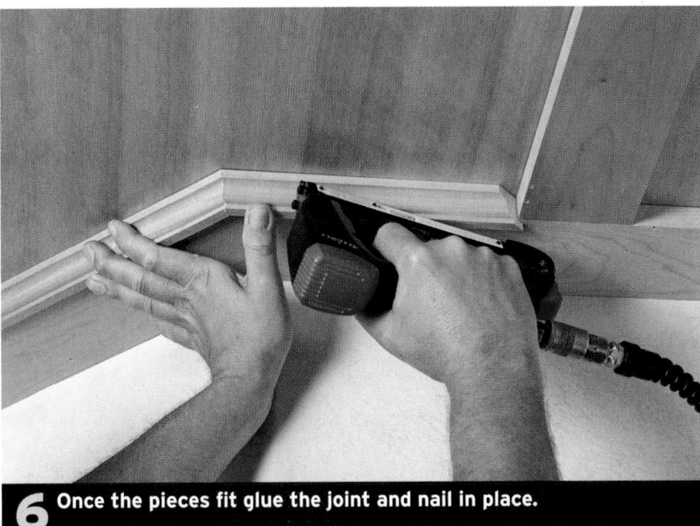

6 Once the pieces fit glue the joint and nail in place.

faster if you take the time to rent and set up commercial scaffolding with locking wheels. An alternative to the heavy commercial scaffolding on ceilings under 10 ft. high is the lighter weight scaffolding sold at most home improvement stores. This scaffolding comes with lockable wheels, which are a great help when moving them around the room.

7 Miter both ends of the closure piece and install.

INSTALLING PRE-ASSEMBLED PANEL MOLDINGS

To make and install pre-assembled rabbeted moldings begin by measuring two adjacent sides of a square panel. Then subtract 1/8 in.from the overall lengths to allow for some fudging, if necessary. Rabbeted moldings are measured from the face of the rabbet not the face of the molding.

Start assembling the pieces by applying glue to both surfaces of the first miter joint. Assemble the miter by shooting a pin nail through the edge and across the corner of the joint ❶; add a spring clamp to tighten the joint further until the glue dries ❷. Repeat this for each joint until the frame is assembled, then set the frame aside until the glue sets up. After the glue has dried, lift the assembled unit in place ❸. Since the moldings are assembled a little undersized it is possible for them to get out of alignment with one another. This makes it important to keep the corners aligned and the sides straight before nailing them in place to the trim ❹.

The finished ceiling, with its mosaic of trim boards and inset panels doesn't just infuse a room with coziness—it lends the entire space a sense of both structural fortitude and traditional elegance, making it feel more secure, formal, and important.

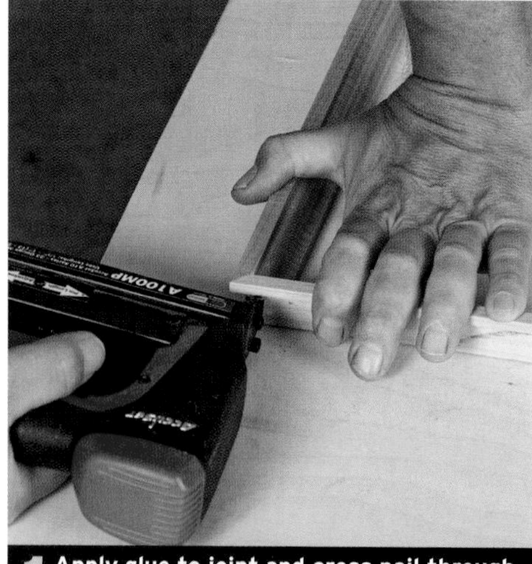

1 Apply glue to joint and cross nail through back of miter joint.

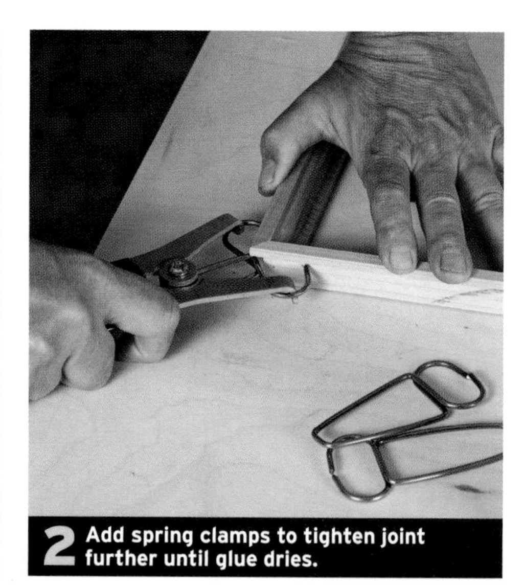

2 Add spring clamps to tighten joint further until glue dries.

3 After glue has dried insert the assembled unit in place.

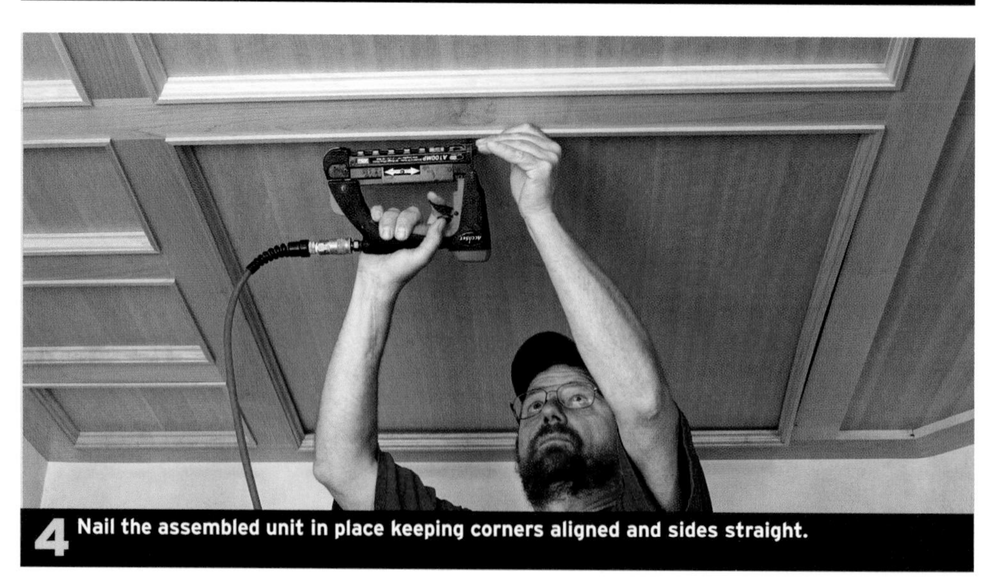

4 Nail the assembled unit in place keeping corners aligned and sides straight.

EXPOSED BEAM CEILINGS

Exposed ceiling beams recall the rugged timber-frame con-struction of traditional buildings. They add historical flavor to period homes, but they can also be used to dramatic effect in more contemporary styles.

Since faux posts and beams are purely decorative, you can lay them out without any regard to the actual structural framing of the building. They may tie into existing wall and ceiling structures or they may just be on their own. Usually, however, beams look more appropriate when they do, in fact, run in the same direction as the internal framing. For example, floor joists would not normally span the long axis of most rooms and neither would timber beams. Unfortunately, to install faux beams in this case, blocking is required between the framing unless you can get away aesthetically with landing the beams directly under the existing framing.

Photo by David Duncan Livingston

Box beams have a more refined look and came into fashion after the rustic colonial period. They can be painted and used with wallpaper to make a distinctive statement.

Photo by Tim Street-Porter

Rustic, widely spaced beams are characteristic of traditional Southwest architecture. These beams complement the other rustic wood used in this room.

Photo by Roe Osborne, courtesy *Fine Homebuilding*

The beams in this timber-framed house are structural. The massive old-growth timbers were cut from lumber salved from a demolished factory.

INSTALLING BEAM NAILERS

1 Use supports to hold nailer in position prior to attaching to ceiling.

2 Screw solid nailer directly into the ceiling joists.

With the exception of flat-applied boards, other types of faux beams are going to require a nailer attached to the ceiling to secure the beams. I use either solid boards or fabricated hollow U-shaped nailers. The location of the beams is laid out on the ceiling with a chalk line, marking out all the beam locations and any intersections. If blocking was installed by the framers between the ceiling joists in anticipation of the beams, nailer installation will be straightforward. If there is no framing to attach the nailers to, then the nailers will have to be attached to the ceiling with butterfly bolts and adhesive.

➔ **See "Attaching Beam Nailers with Butterfly Bolts," on the facing page.**

After verifying where the blocking is by measurement (or, lacking measurements, by the use of a stud finder or nail probe), measure from the adjacent walls and chalk lines for both sides of the nailers.

Prior to installing the beam nailers, draw the entire beam layout including any perimeter beams directly onto the ceiling Sheetrock. This provides a full size preview of the layout as well as an opportunity to catch any measurement mistakes before actually installing anything to the ceiling. Mark the locations of the nailers and use a chalk line to draw both sides of the nailers onto the ceiling. Once the layout is transferred to the ceiling, fasten the nailers to the ceiling. The two types of nailers are fastened to the ceiling using the same techniques.

For good results, solid nailers need to be straight, free of twist, and dry. When working alone it helps to use adjustable or fixed supports to hold the nailer to the ceiling while aligning it to the chalk line prior to attaching it to the ceiling ❶. Whenever possible, screw the nailer directly into the ceiling framing ❷. If there is no framing available, the nailer is backed with construction adhesive and fastened to the Sheetrock using butterfly bolts. Solid nailers are generally run for the full length of a beam.

Hollow nailers are assembled before installation. Start building the nailers by ripping all the nailer stock to size and cutting the pieces into 1-ft. lengths. The width of the top board is determined by subtracting the combined thickness of the sides from the overall width of the hollow box. Assemble the individual pieces by nailing or screwing the sides into the edge of the top board ❸. Align the top edges of the box nailers between the chalked layout lines and fasten them to the ceiling either with screws into the ceiling framing or use butterfly bolts and adhesive on the Sheetrock ❹. Place the boxes at each end of the beam and every 4 ft. in the middle of the opening.

3 Assemble the sections of box nailer.

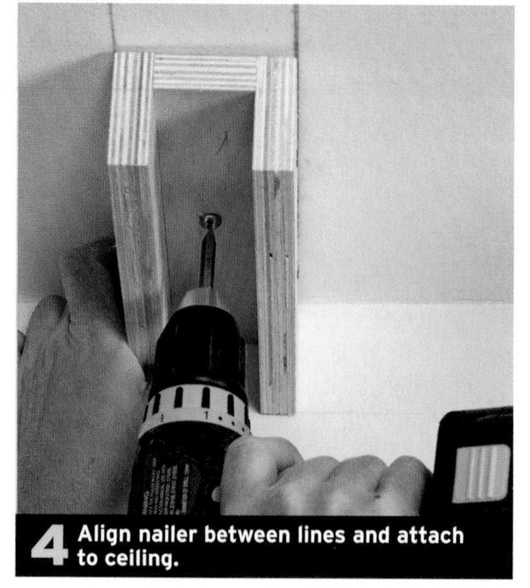

4 Align nailer between lines and attach to ceiling.

ATTACHING BEAM NAILERS WITH BUTTERFLY BOLTS

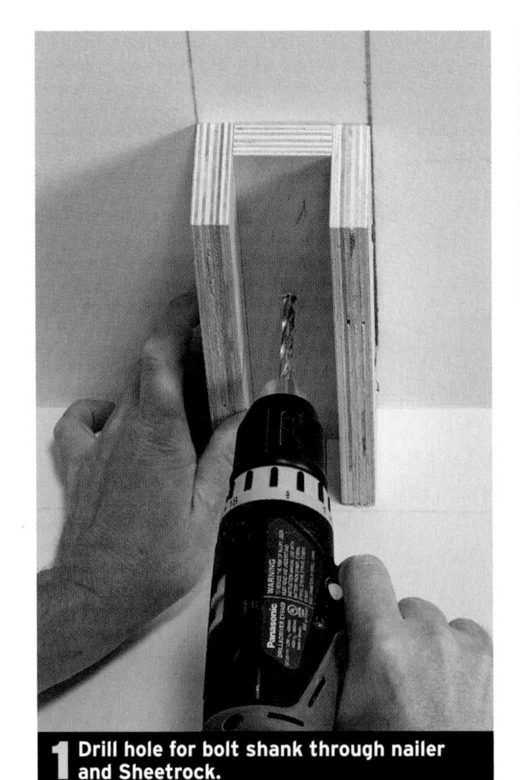

1 Drill hole for bolt shank through nailer and Sheetrock.

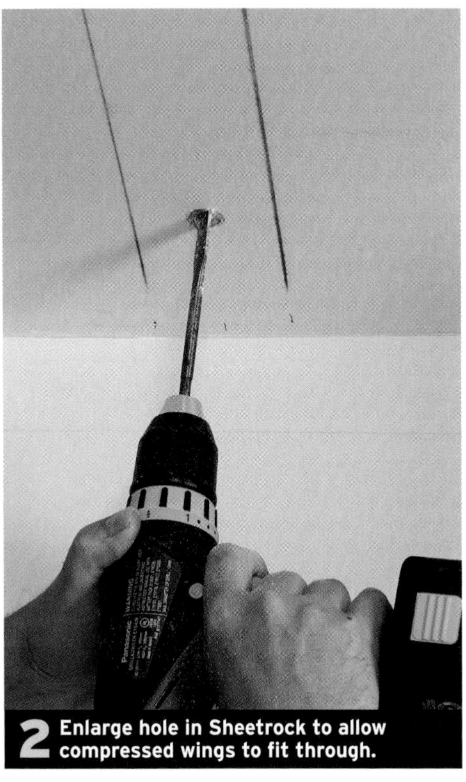

2 Enlarge hole in Sheetrock to allow compressed wings to fit through.

3 Push the wing assembly through the hole in the ceiling Sheetrock.

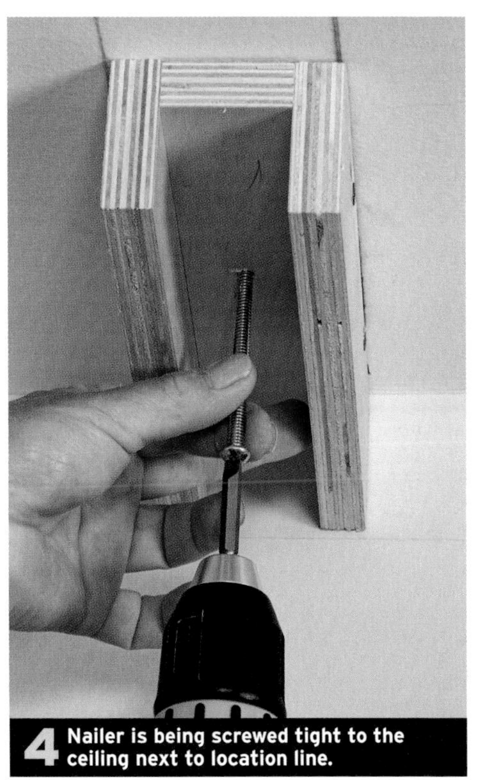

4 Nailer is being screwed tight to the ceiling next to location line.

Butterfly fasteners provide a secure way to attach a nailer to Sheetrock and are strong enough to compress the construction adhesive. Regardless if the nailer is a full-length solid nailer or a short box nailer there should be nailer and fastener placed at each end and every 3 ft. or so along the length of the beam.

Start by holding the nailer in position and drill a 3/8-in. hole through the nailer and the Sheetrock ❶. Then use a spade bit to enlarge the hole in the Sheetrock enough to provide clearance for the compressed wings ❷. Insert the bolt through the predrilled hole in the nailer and screw on the wing assembly. Now apply construction adhesive to the back of the nailer, compress the wings, and slip them through the hole in the ceiling ❸. Pull down slightly on the bolt while tightening to keep the wing assembly from spinning ❹.

FAUX BEAM OVER HOLLOW NAILERS

I use hollow box nailers when working alone because they are relatively small, lightweight, and easy to handle. The box nailer sides are ripped to the correct depth and serve as the depth stops for the bottom beam cap. Other than the nailers themselves the order of assembly of the beam parts is the only difference in installation. (See the drawing at right.) As the nailers are attached by nailing through their relatively thin edges, it is important that the material resists splitting when a fastener is driven into it. This means MDF is not a good option. The box nailers are assembled in short sections and installed next to all the beam intersections and to every other joist.

If the beam runs parallel and between the joists install the box sections every 4 ft. using adhesive and butterfly fasteners through the Sheetrock.

Allow the nailer adhesive to set for a day and then begin by installing the bottom cap board. Make a spiling batten from a 6-ft.-long strip of door skin or other thin plywood.

TRADE SECRET

If the beam sideboard is over 10 ft., add 1/8 in. to the overall length. You can then bend and spring the board in place, ensuring a tight fit at both ends.

Align the batten to the bottom edge of the nailer and butt one end against the wall or perimeter beam. If the joint angle is not square, scribe and cut the batten until it fits ❶. Repeat this for the opposite wall using the other end of the batten. Now determine the length needed by using a tape measure or pinch sticks ❷ and transfer both the length and the batten end cuts onto the bottom cap board ❸. Cut both ends of the board on the traced lines and hold it in place to check for fit. If everything looks good, nail the beam cap to the bottom of the nailer box sides.

To ensure a tight fit between the finished sides and the bottom cap, be careful to keep the edges of the nailer flush to the edges of the bottom cap ❹. Repeat the fitting and cutting process for the beam sides. When installing the beam side, push the top edge up to the ceiling as tightly as it will go and check for gaps. If the gaps are small they can be filled with caulking. Gaps larger than 1/16 in. should be eliminated through scribing and trimming as necessary with a hand plane.

After the sideboards have been nailed to the nailers ❺, nail them to the bottom cap using a clamp to draw the boards together if necessary as the nailing progresses down the board ❻. Larger gaps may require scribing and cutting with a jig or circular saw. Alternatively, you can cover the gaps with a molding ❼. The finished installation ❽.

1 Scribe spiling batten to wall surface.

BOX BEAMS

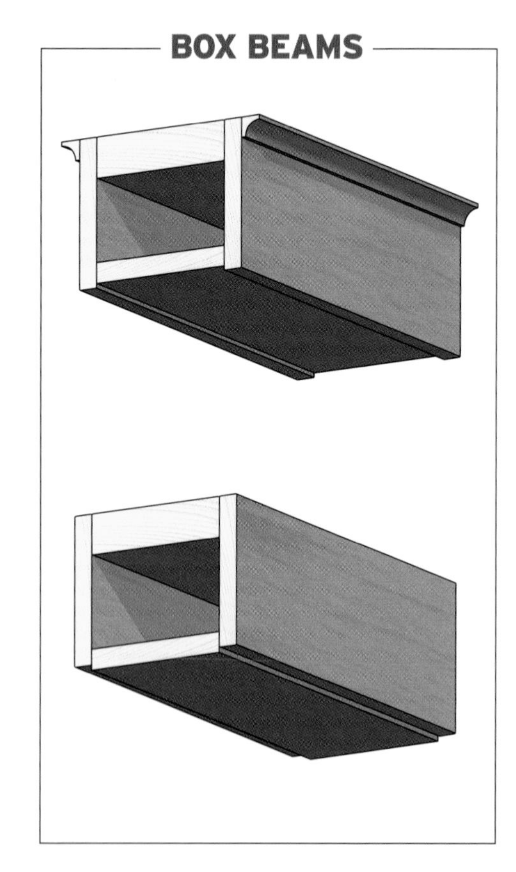

2 Use pinchsticks to determine overall length.

3 Use batten to transfer end angle to bottom cap.

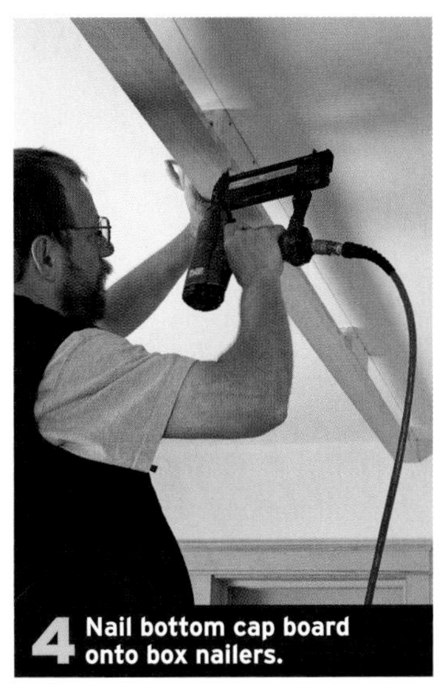

4 Nail bottom cap board onto box nailers.

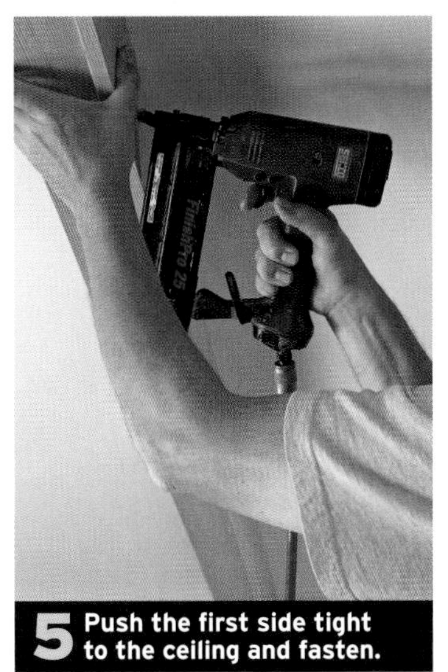

5 Push the first side tight to the ceiling and fasten.

6 Use clamps to draw sides tight to bottom cap while nailing.

7 Small molding being applied to ceiling/beam intersection.

8 Finished faux beams.

FAUX BEAM OVER SOLID NAILERS

Once you've gotten the nailers installed, start the faux beam construction by cutting and installing the sidepieces with their preinstalled stops. The sides of the beam are installed prior to installing the bottom cap. Then insert the bottom beam cap between the two sides against the stops and clamp it in place for nailing.

If there are perimeter beams they will need to be installed first. To avoid gaps where the sides of the beams meet the walls (unless you intend to cover the butt ends with a running molding) you will likely have to scribe them at each end. This also holds true for where the top edge of the beam sides joins with the ceiling. You can avoid scribing by installing a running molding to cover the joints—a faster process than scribing the boards to the ceiling. Even if you don't need to scribe, moldings are an elegant option as they add shadow lines that give the beams an increased feeling of depth and substance.

Start by using a spiling batten and pinch sticks to fit and cut the sides to length.

Next nail the bottom cap stop strips to the inside faces of the sideboards using a gauge block to position the stop strip while nailing ❶. Once the sides are cut to length and the stops applied, fasten them to the ceiling nailer. Be sure to keep the top edges pressed tight to the ceiling ❷. If necessary, scribe the top edge to fit or alternately use an additional molding at the top. Insert the bottom cap between the two sides and push it up tight against the stops ❸. Use quick clamps to help draw the sides tight and nail the sides into the edge of the bottom cap ❹.

1 Use locator block to position stop strip for nailing.

2 Nail the beam sides to the solid nailer.

3 Insert the bottom cap.

4 Clamp sides tight and nail cap to bottom.

FAUX POSTS

design options

Photo by Rob Karosis

These posts in a bungalow living room give the illusion of frame and panel, created by the application of corner trim.

Photo by Grey Crawford

Used in a colonial setting, these posts not only tie the trimwork together, they provide an effective division of a large space into smaller, more intimate spaces.

Adding faux posts to an exposed beam ceiling give the appearance of a true post-and-beam frame in which exposed posts carry the weight of the overlying structure to the foundation. Use one of three ways to build a non-structural post or to cover an existing post with boards. The simplest is a post against a wall is to edge nail three boards around a nailer—a typical strategy used in exterior work where the joints will be caulked and the post painted. A freestanding structural post is covered on all four sides.

Another method—boards trimmed out with a corner molding—isn't really any more difficult to do, though it does add more pieces to cut and install. This method avoids having to nail the individual sides to one another so fitting the boards is more forgiving. The moldings also create another level of appealing, substantive detail. Either method requires you to install nailers (sized to the interior width, though not necessarily the depth, of the faux post) where you wish the posts to appear.

Photo by Rob Karosis

This simple faced post provides a focal point in a room with a vaulted ceiling. In this case the post is structural, but facing it allows the post to become a design element.

TRIMMING A FREESTANDING POST

1 The first pair of boards are being nailed on to a built-up 2x4 on a standing post.

3 Use clamps to draw joint tight while nailing sides together.

2 Check with a straightedge to ensure edges extend past rough post.

Start by milling the trim stock to the correct dimensions: A few inches over length and at least 1⁄16 in. over the width of the existing post. (If the post is rough lumber, be sure to find the widest points.) The second two opposing boards will be wider than the first two by twice the thickness of the boards. Joint the trim stock straight with edges square.

Cut the two opposing trim boards to length and nail the boards to the post ❶. Be sure that the edges of the trim boards extend past the two opposing faces of the existing post or nailer ❷. Now install the second pair of wider boards by flushing their faces to the outside edges of the installed trim boards. To help with alignment and close up any gaps, set clamps around the area you are nailing ❸.

FAUX POST WITH CORNER MOLDINGS

Trimming out a faux post with a corner molding isn't really any more difficult to do, though it does add more pieces to cut and install. This method avoids having to nail the individual sides to one another so fitting the boards is more forgiving. The moldings also create another level of appealing, substantive detail. Install nailers (sized to the interior width, though not necessarily the depth, of the faux post) where you wish the posts to appear.

Install the trim boards to the faces of the post/nailer and then create corners with a length of quarter molding. Begin by milling the trim facing boards to the same widths as the exposed sides of the nailer. Fasten the two opposing sides onto the nailer being sure to keep their edges flush with the nailer's face ❶. Next, install the third trim board (and fourth if the post is freestanding and not against a wall), lining it up with the two inside edges of the installed facing boards ❷. Now rip and shape the corner moldings (or use a stock molding such as quarter-round) to a dimension that will leave about a 3/16-in. reveal and nail them into the inside corners created between the sides and the face board ❸.

1 Keep edges flush to rough post when attaching first two opposite sides.

2 Attach remaining two sides keeping them flush with edges of rough post.

3 Leave a 3/16-in. reveal when attaching the corner pieces.

ADVANCED TECHNIQUES & PROJECTS

THIS SECTION IS ABOUT PUTTING SKILLS and techniques together to create something special. Many so-called advanced projects are within the reach of anyone who feels comfortable with hand and power tools. For example, installing complex crown molding is a matter of good planning and accurate layout, but uses the same techniques as installing a simple crown.

Sometimes a project calls for something more than off-the-shelf stock moldings. You can create unique and dramatic effects by combining moldings to make more complex moldings. The combinations you can make are nearly limitless, especially if you make your own moldings. This comes in handy when you want to make naturally finished moldings of a particular species of wood. Fortunately, it's easy to create custom moldings with a router, a router table, and a tablesaw.

CUSTOM MOLDINGS

RADIUSED MOLDINGS

ADVANCED PROJECTS

THOMAS

COMBINING MOLDINGS

Designing a larger molding by combining different individual stock moldings is an alternative to buying a large single-piece molding. A combined molding will typically feature more pronounced detail than a similar-shaped commercial one-piece molding. Due to their size some trim elements such as cornices must be built up out of various parts. Other components such as casings and base—and even the crown molding used in a cornice assembly—can be built by combining different smaller moldings.

The single most important factor to consider when creating a combined molding is its proportion—both between its own elements and to the room it will appear in. Typically, the larger the room the larger the trim should be. For example, a tall ceiling supports a large cornice molding, but a simpler and smaller two- or three-piece combination works well in an 8 ft. or 9 ft. high room. The two-piece molding **A** will work in rooms 8 ft. tall as long as the size of the crown molding is kept smaller than 4 in. or so. Adding another piece of base to the ceiling turns this into a three-piece cornice **B** and adds a bit more mass to help the overall look as the room grows higher. If you wanted to produce a similar effect using a router table (rather than a shaper capable of larger profiles), the easiest way would be to combine smaller pieces to create combination moldings similar to the examples in the drawings at right **C**, **D**.

Casings and baseboard are generally not made of more than two or three separate elements. Back banding a casing is a good way to give the casing more depth.

➔ **See "Back Band and Casing Combination," p. 214.**

Adding a base cap and a shoe molding does the same thing for baseboard. Be conservative when combining elements and look to traditional examples for inspiration. The previous generations of craftsmen have had it figured out for a long time now.

CROWN OPTIONS

The number of elements and how they are placed affect the appearance of crown molding.

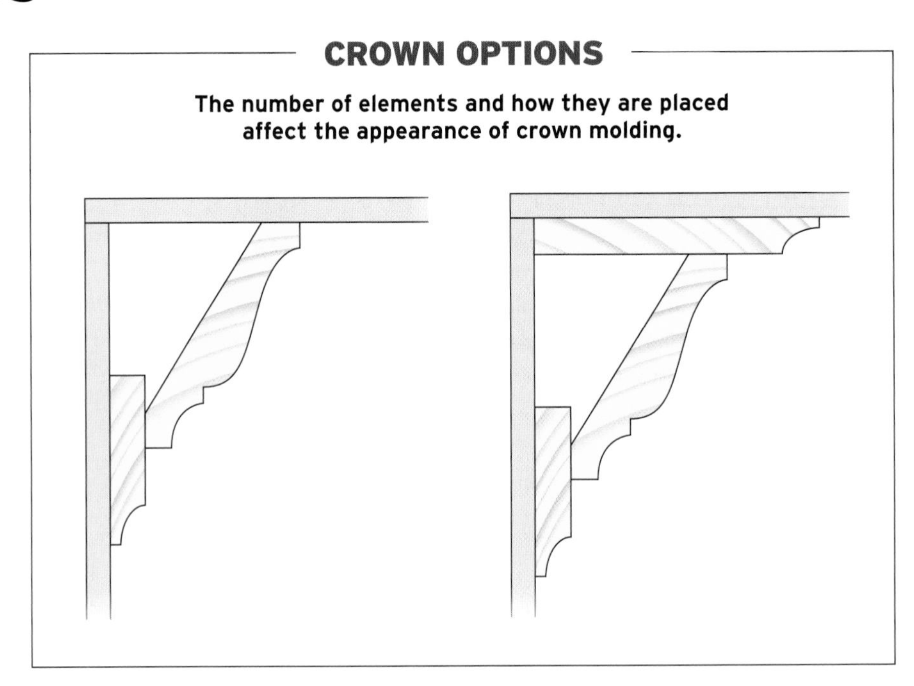

ROUTER CUT CORNICE MOLDINGS

Dotted line indicates standard cornice molding.

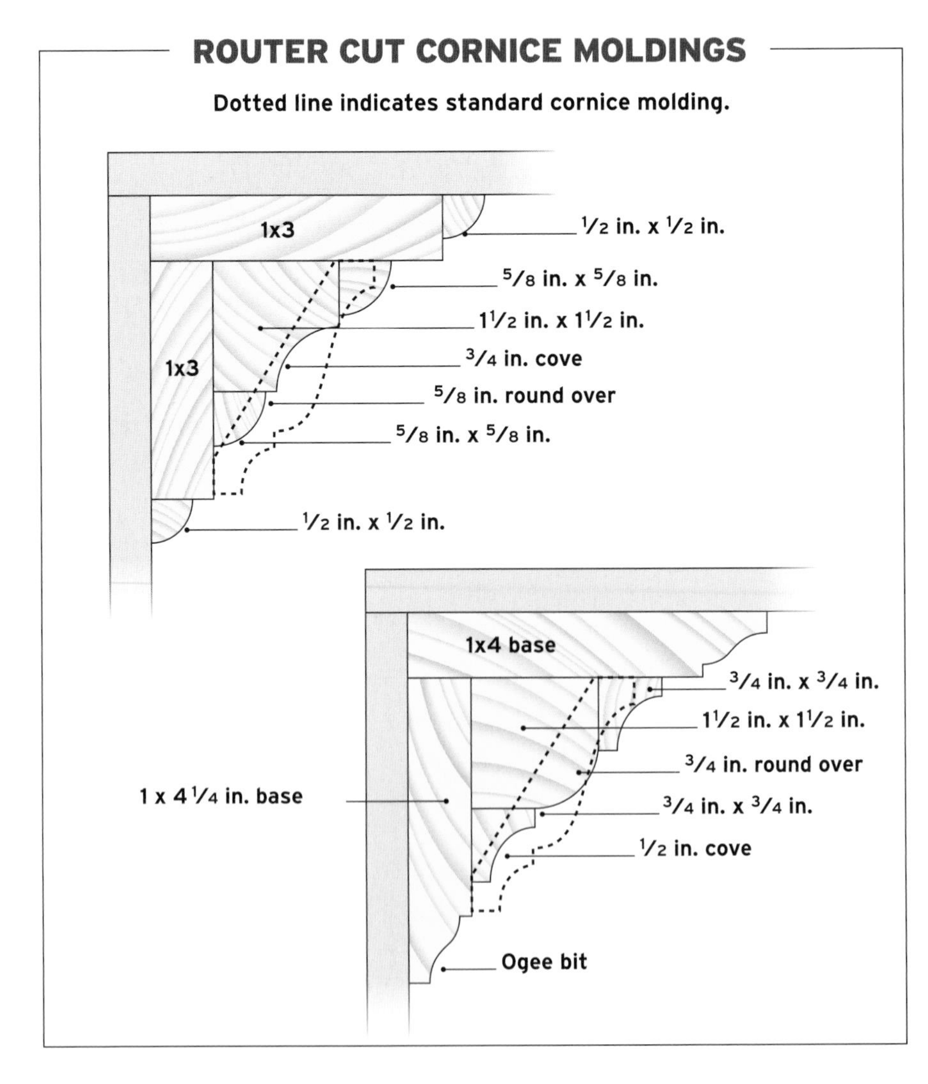

SIMPLE MOLDING ON THE ROUTER TABLE

A surprising variety of trim profiles can be made by using a router mounted in a router table and a tablesaw. A router of around two horsepower will be needed as well as a tablesaw. The router table should have an adjustable fence and dust collection. When using larger bits, make multiple passes to avoid tearing the material or overheating the motor. A slow even feed rate is best and will provide the smoothest finish.

Begin by mounting a bit of the desired profile in the router. The bit shown here is an ogee. Set the height of the bit to the cutting depth of the profile ❶. Setting different heights will give slightly different profiles. Also, running the board on the edge with the face run against the fence versus running the edge against the fence ❷.

Now use a straightedge to set the near side of the bearing flush to the face of the router table fence ❸. Feed the stock smoothly from right to left, holding it firmly to the fence and the table ❹. Set the tablesaw ripped fence to the desired width of the molding. Using a push stick, rip the molding free of the board ❺.

WARNING

Whenever making small moldings on the router table use wide enough boards to keep your hands out of harm's way and use featherboards whenever cutting narrow stock.

1 **Set the height of the bit so that it cuts the desired depth.**

2 **Cutting the profile on the edge or the face gives you different profiles.**

3 **Use a straightedge to align the edge of the bearing with the fence.**

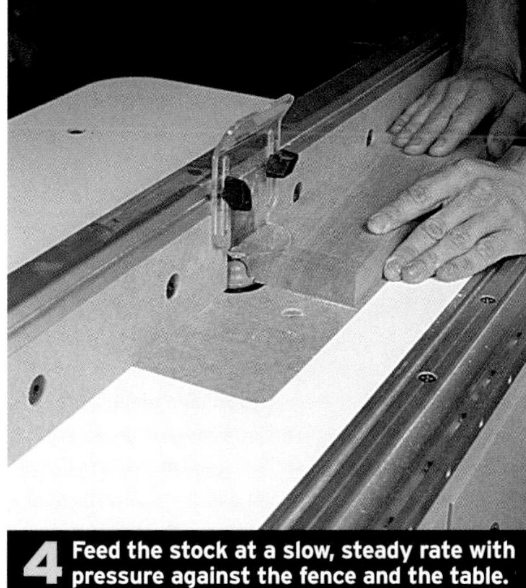

4 **Feed the stock at a slow, steady rate with pressure against the fence and the table.**

5 **Cut the molding away from the stock on the tablesaw.**

Photos by Lonnie Bird

TRADE SECRET

When making moldings make more than you actually need. It avoids having to set up the router table again if you make a mistake and run short. Keeping a sample helps reset the height if you need more.

BUILT-UP MOLDING

You can make very elaborate moldings by stacking separate strips. In the example, we are stacking the boards onto their face, but you can also stack moldings on edge, using the basic shapes to create unique profiles.

See "Basic Molding Profiles," p. 10.

In this example, we're using a molding cutter adjusted to different positions to rout several profiles.

Begin by mounting the bit in the router and tightening the collet securely ❶. Set the height of the bit to rout the first profile ❷. Because this is a large bit, you'll have to take at least two passes. Adjust the fence for a light pass. Minimize the amount of cutter exposed by the fence. If you have an adjustable fence, bring the panels closer to the bit ❸. Feed at a steady rate and keep the stock tight to the fence and table ❹. Depending on the hp of the router, you may need to take several light passes. If the motor is straining or there is burning, back off and take lighter passes. Then, align the bearing with the fence ❺. Rout the full depth of the profile, holding the stock firmly to the fence and the table ❻. To make the cove in the top molding, raise the bit to the desired height ❼. Position the fence. In this case, we are setting it tangent to the fillet profile below the cove ❽. Rout the cove cut ❾. The finished molding ❿.

1 Inset the bit in the collet and tighten it securely.

Molding bits create complex profiles in one pass.

WARNING When using large bits, adjust the speed of the router to the bit manufacturer's recommendations. See the chart below for general guidelines. Large bits should be run at lower speeds for safety.

Complex molding bits can create many profiles by adjusting the height of the bit and routing on the edge or face of the board.

Photos by Lonnie Bird

SUGGESTED ROUTER SPEEDS

BIT DIAMETER	MAXIMUM SPEED
Up to 1 in.	24,000 RPM
1 in. to 2 in.	18,000 RPM
2 in. to 3 in.	16,000 RPM
Over 3 in.	12,000 RPM

2 Set the height of the profile.

3 Minimize the fence opening.

4 Make light passes until nearly at full depth.

5 Use a straightedge to align the bearing with the fence.

6 Rout the full depth profile holding the work firmly to the fence and table.

7 For the second piece, adjust the height of the bit.

8 Adjust the depth of the cut.

9 Rout the second profile.

10 The finished molding.

Photos by Lonnie Bird

COMPLEX MOLDING WITH MULTIPLE BITS

1 Remove only half of the material on the initial cut.

2 Reset the router depth higher and finish cutting the cove profile.

Multi-profile router bits are expensive and their large diameter means you need a variable speed router with at least two horsepower. But you can also make unique moldings using several bits. This molding requires several router bits to produce. After setting the router and the fence to the correct position start by shaping the cove. Feed the board is fed on edge. This profile is too deep to cut in one pass, so set the router bit for the initial cut to remove only half the finished depth ❶.

WARNING Trying to remove too much material at once is unsafe and produces a ragged, sometimes burned surface.

Raise the height of the router to its proper height and run the stock through a second time ❷.

Now remove the cove bit from the router collet and switch to the roundover bit. After resetting the fence and the cutter to the correct positions rotate the stock ninety degrees and lay it on its face to shape the roundover ❸. Once the roundover is completed rip the molding free from the board ❹.

TRADE SECRET In a multiple step process like this, be sure you cut all your pieces of molding at each stage so you don't have go back and attempt to reset the bit.

3 Install the roundover bit and after resetting the router and fence rotate the stock ninety degrees and finish the profile.

4 Cut the molding free on the table saw.

Photos by Lonnie Bird

MAKING DENTIL MOLDINGS

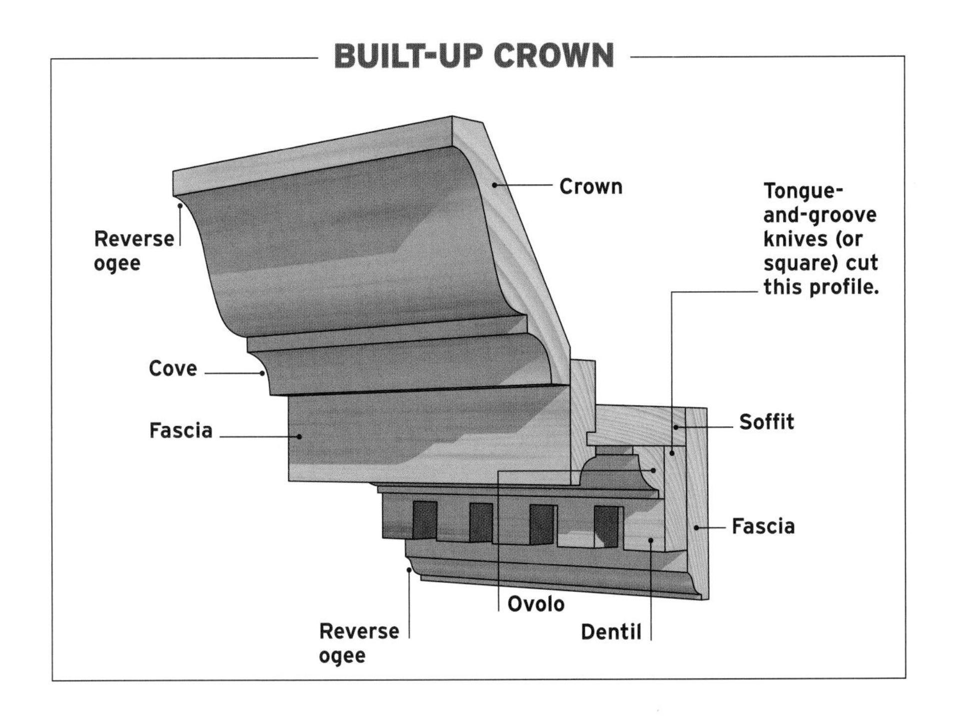

Dentil moldings can make a dramatic base to a cornice molding.

The easiest way to make dentil is on a tablesaw using a dado head. If you don't have a dado head, you can cut the dentils by cutting away on the inside of each mark and then nibbling away the remaining stock in successive passes.

Work out the proportions and draw it to scale beforehand. Use simple layout tools to draw the molding layout on a piece of the stock you will use to cut the dentil.

Install an auxillary to the tablesaws miter gauge by screwing a piece of flat lumber to the miter gauge. This provides additional support and creates a surface to accept a stop. Now stack the dado to the width of the desired kerf in the dentil, set the blade to the correct depth, and then make a cut through the backboard. Transfer the dentil spacing to the backboard with a set of dividers and pound a small brad into the backboard at that point leaving a bit of the nail protruding to act as a stop ❶. Push the end of the molding material against the stop and make the first cut ❷. To maintain the spacing of the cuts, index the previous cut over the brad ❸.

1 Mark the spacing on the backboard driving a brad at that point.

2 Make the first cut using the brad as a stop for the butted molding.

3 Continue to use the brad for a stop on the additional cuts.

Photos by Lonnie Bird

BACK BAND AND CASING COMBINATION

1 Rout molded profile on edge of back band material.

2 Use the rabbet outline to set table saw blade height and fence.

3 Cut the waste from the rabbet.

4 Rout molded profile on edge of the casing material.

5 This combination produces a simple but elegant back band and casing.

This casing and back band combination is made from standard 3/4 in. boards using a roman ogee bit in a router table and a tablesaw. Start by creating the back band: First rout the ogee profile on the edge of the 3/4-in. back band material by laying the material flat on the table with its edge against the router table fence ❶. Now use a square to draw the outline of the rabbet on the end of the stock. Use the drawn outline to set both the blade depth and the fence setting on the tablesaw for each of the cuts required to make the rabbet ❷.

Make both the rabbet cuts, making sure the waste is on the outside, away from the fence ❸. Now make the casing. Leave the router bit in the same position and run the casing through it with its face flat on the table ❹. The completed trim combination is simple but elegant ❺.

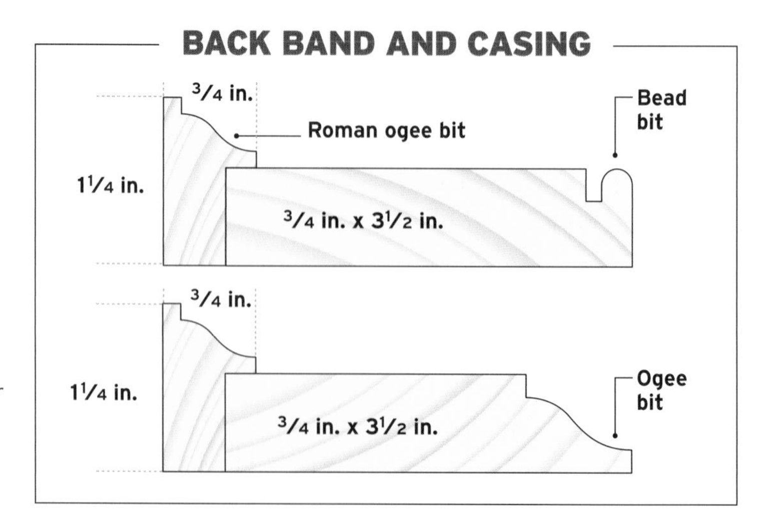

FLUTING WITH A ROUTER

While you can buy fluted molding in many homecenters or order it from a custom mill shop, it's easy to cut with a router. You'll need an edge guide, either a commercial accessory or make a jig such as shown here.

The jig is made using a piece of 1/2 in. plywood wider than the stock to be fluted, a piece of straight stock for a fence, and two carriage bolts and wing nuts. Cut two slots, wide enough for the bolt to pass through, approximately 1 in. from each end of the plywood. Tightening and loosening the wing nuts allows you to set the distance of the fence. Drill a hole in the base large enough for a core box bit to pass through and drill holes so that you can attach the jig to the base of your router ❶.

Set the router bit to the desired depth of the flute ❷. Lay out the flutes so that they are the same distance from one another. Cut the edge of a sample board to use as a gauge for setting the fence for the actual cuts ❸. Determine the exact length of the flutes, allowing for the unfluted stock at the top and bottom. Clamp stock blocks on each end of the stock to stop the cut in the same location as each flute is routed ❹. Cut the center flutes first and work outward ❺.

Holding the bit against the fence to begin each cut, carefully lower the bit into the stock and cut the remaining flutes ❻.

1 Make an edge-routing jig and attach it to the router or use a commercial edge guide.

2 Set the depth of a core box bit to the depth of the flute.

3 Carefully layout the flutes equidistant from one another. Rout a sample board to facilitate layout.

4 To begin the cut, set the base against the stop block and lower the bit into the work.

5 Rout the center flutes first and work to the outside.

6 Cut the outside flutes.

Photos by Lonnie Bird

CUTTING COVES ON THE TABLESAW

1 Mark the height and width of the cove and raise the blade to the depth mark.

2 Move the fence until the blade will enter on the far width mark.

3 Check to see that the blade aligns with the near width mark as it exits.

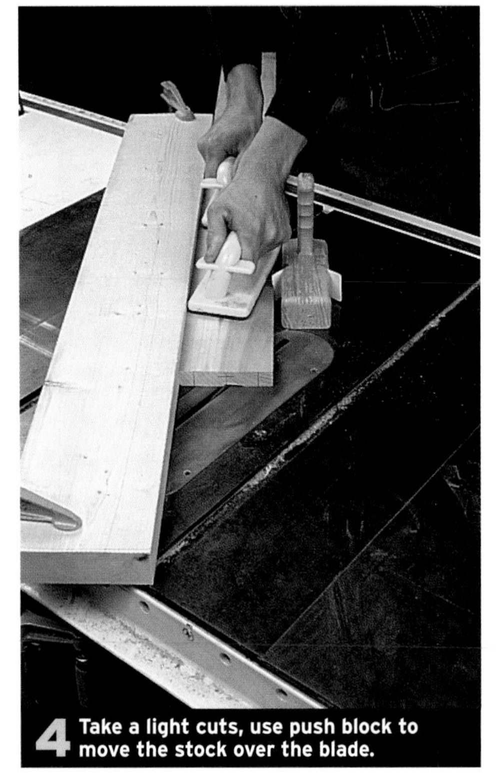

4 Take a light cuts, use push block to move the stock over the blade.

5 Take successful passes raising the blade 1⁄16 in. for each cut.

6 The finished cove can be sanded smooth.

Photos by Lonnie Bird

A cove shape is the foundation for many crown moldings. It can be cut on the tablesaw by clamping a fence at an angle to the blade. The drawing at right shows how different angles of the fence to the blade enable you to cut differently shaped coves. Begin by marking the width and height of the cove on both ends of a test piece, the same width and thickness as the stock you'll be coving. Adjust the height of the blade to mark, which indicates the depth of the cove ❶. To position the fence, set it so that it will enter the stock on one edge of the cove ❷ and exit it on the other end of the cove ❸. Clamp the fence tightly at both the front and back of the tablesaw.

WARNING Do not attempt this technique unless you can securely clamp the fence.

Lower the blade to take a light cut. Using push blocks to keep your hands away from the blade, hold the stock firmly to the fence and push it over the blade ❹. Raise the blade in small increments (about 1/16 in. with each pass) until the cove is cut to the depth mark ❺. The finished cove can be sanded or scraped with a radiused hand scraper until smooth ❻.

Safety guidelines for cutting coves

- Clamp the fence securely on both ends
- Hold the work tight to the fence with a featherboard
- Take light cuts
- Use push blocks

FENCE ANGLES FOR CUTTING COVES

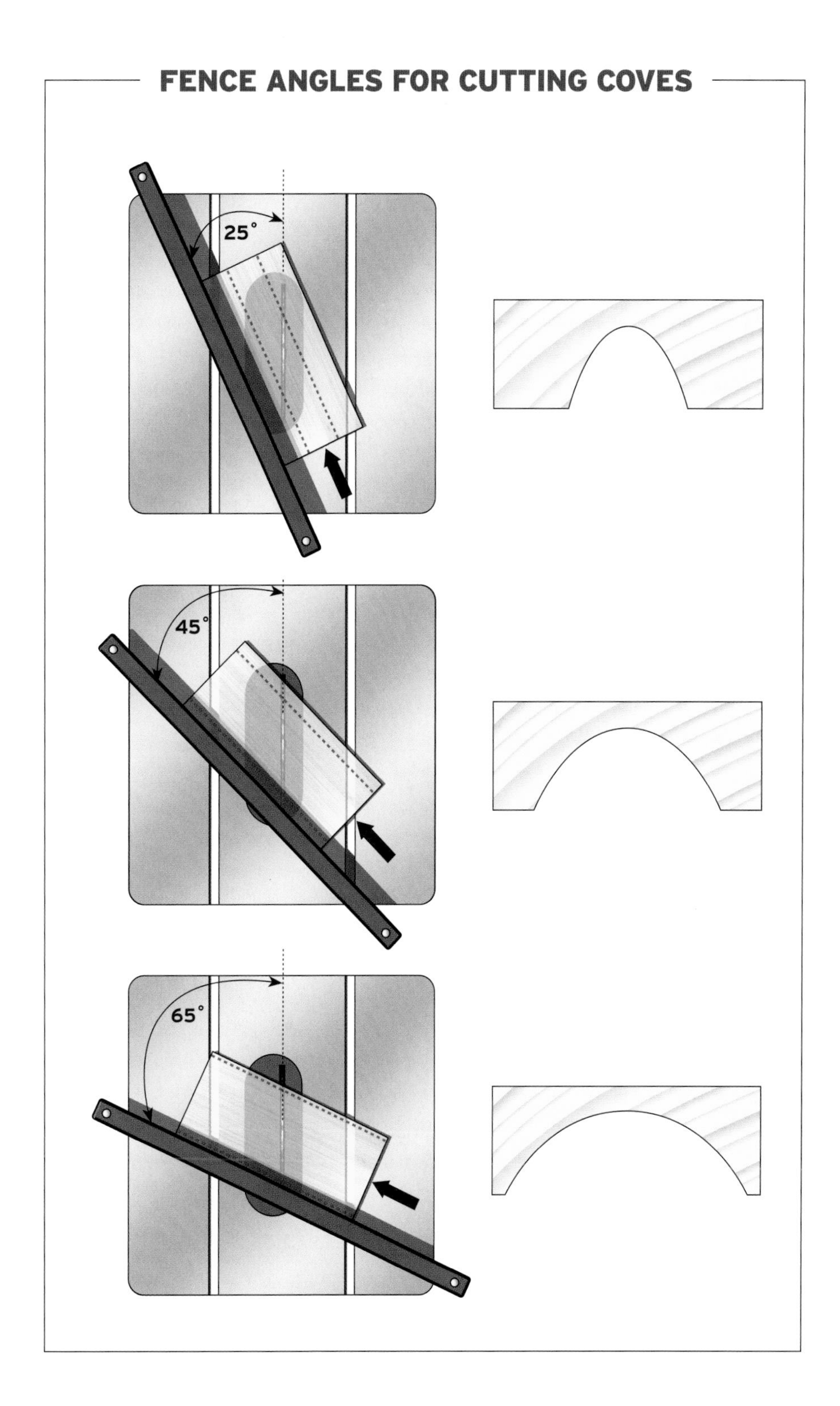

BENDING MILLWORK

When I work with paint-grade materials, I use bendable urethane moldings as they are easier to deal with than wood moldings. If, however, the moldings are to be naturally finished wood they will need to be bent or laminated in thin strips to fit the radius of the bend.

There are three ways to bend a molding around a curve: kerfing, laminating, or steaming. Of the three, kerfing is the fastest and easiest—though since the kerfs will show it is not appropriate if the edge of the board is going to be exposed.

Different woods and grain configurations bend differently so take the time to experiment a bit before cutting and bending the actual pieces. Due to the difficulty of getting an exact measurement around a curve it is best to install the curved portion first and then fit the adjoining pieces to it. If this is not possible, cut the piece to be curved long and then butt one end to the finished millwork on one end, wrap it around the curve, and then make a cut mark where it overlaps the molding at the other end.

While laminating and steaming can sometimes be done in place, it is generally easier to use a bending fixture to represent the curved surface. The fixture is simply a surface to which individual clamping jigs are secured along the desired curve. Alternately, you could stack multiple pieces of plywood cut to the appropriate radius to create a form.

The bending jigs shown in the drawing at right are made from ½ in. plywood and are screwed or, better, bolted in place.

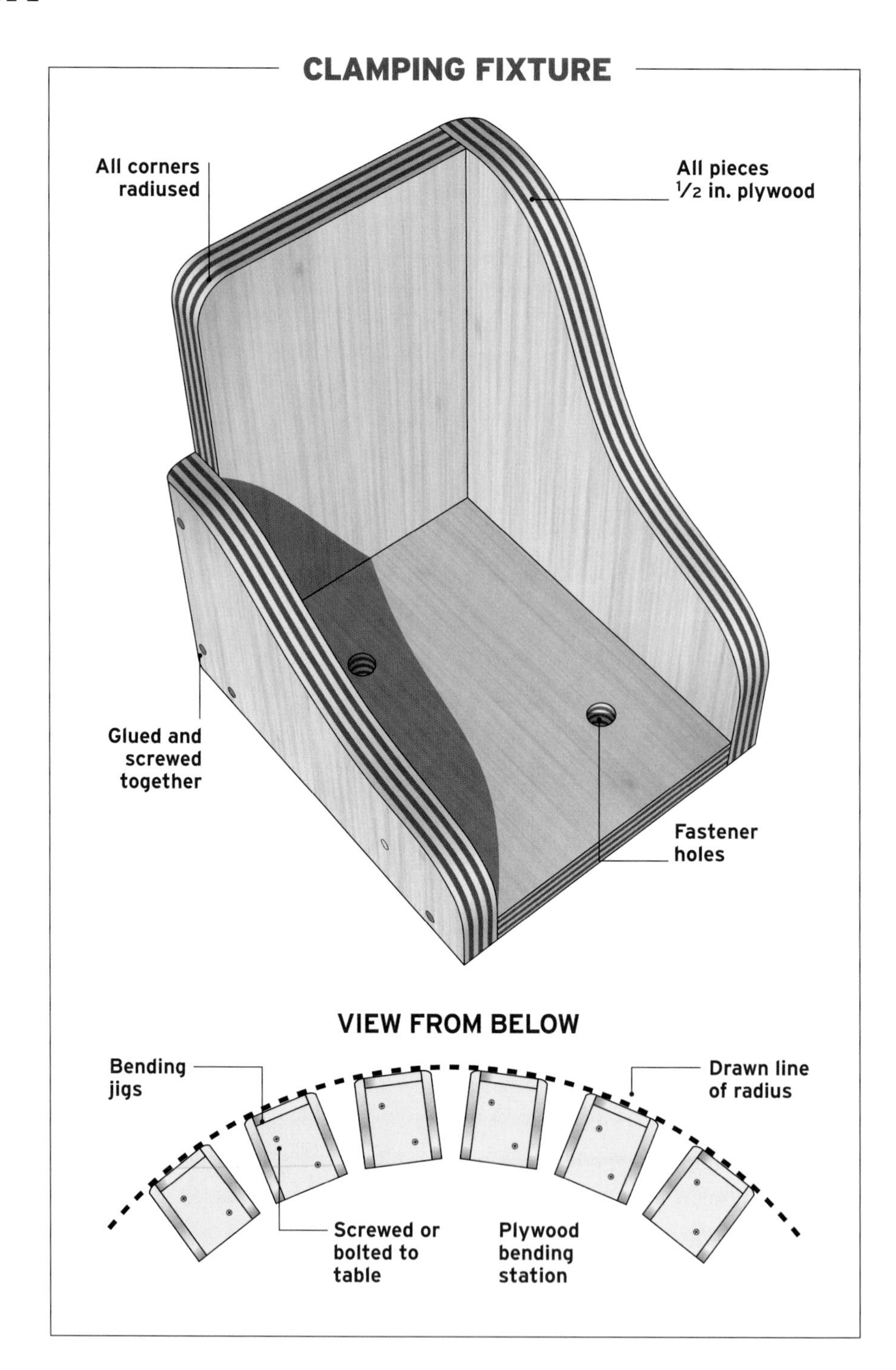

LAMINATING A CURVED MOLDING

Thin laminations of wood, glued together and clamped around a form until the glue dries, forms an accurate, stable, and strong curved molding. Start by determining the thickness and number of laminations that will be required. This is a matter of trial and error, as bendability depends on the thickness and the species of wood as well as the tightness of the bending radius. Be prepared to break a few laminations before finding the correct thickness ❶. When stacking up the laminations, orient the grain in the same direction. This minimizes tear out when planing the edge smooth once the glue is dry. (Note that the clamping stations will almost touch on tight radiuses and get farther apart the larger the radius.) Apply epoxy or yellow wood glue to the mating surfaces of the laminations ❷ and then hold the laminations temporarily together with several wraps of blue masking tape ❸. Start at one end, clamp the lamination to the form making sure to use clamp pads to avoid marring the wood ❹. Continue applying clamps at each station along the radius until the laminations are tight to each other and to the form ❺. After the glue has dried remove the piece and trim it to fit.

1 Determine lamination thickness by trial bending around form.

2 Apply glue to all the mating surfaces.

3 Temporarily tape the laminations together.

4 Clamp laminations at one end and work your way around the curve.

5 There should be at least one clamp at each station.

KERF BENDING A CURVED MOLDING

1 Attach a backboard to the miter fence and cut its end off.

2 Set the saw blade to leave about 1⁄16 in. uncut on board.

Kerfing (closely spaced cuts made into the backside of a board) is an appropriate choice for bending a molding when the kerfed edge will be covered with another molding or another type of woodwork. The tablesaw is the tool of choice for making the multiple, evenly spaced cuts. When determining the length of the curved molding, be sure to include a short portion of any straight wall on either side of the bend in the overall length of the board. This straight portion will be used for nailing the trim to the wall that borders the curved area.

Start by attaching a backboard to the saw's miter fence and then set the miter fence to cut at 90 degrees. Cut the end of the backboard to length by running it through the saw ❶. A smooth bend requires consistent depth and spacing of the kerfs, though the depth and spacing of the cuts is entirely dependent on the material and the bending radius—be prepared to experiment a bit to get the right combination. Now set the kerf cut depth (always leave at least 1⁄16 in. of the board uncut) ❷. Put the molding stock against the backboard of the miter fence and make the first cut ❸. When the blade stops turning, slide the material over to the kerf spacing as determined by your experimentation ❹. Leave the board in this position and draw a line on the table to represent the edge of the previous kerf ❺. Cut the rest of the kerfs by aligning the same edge of the previously cut kerf to the line on the tablesaw ❻. After the board has been kerfed, test its ability to bend to the correct radius before installing ❼.

Since a kerfed board has lost nearly all its strength, handle it with care during installation. Apply construction adhesive to the back side and then install the board to the wall—press to force the adhesive into the kerfs to add strength to the board.

3 Make the first cut through the molding stock.

4 Slide the material over the desired spacing for kerf cuts.

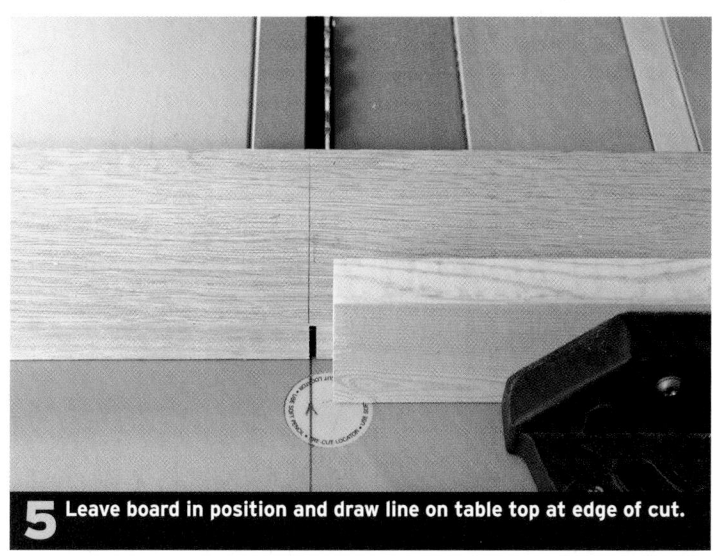
5 Leave board in position and draw line on table top at edge of cut.

6 Align same side of previously cut kerf to line on tablesaw.

7 Finished molding should easily bend around radius.

WHAT CAN GO WRONG

When installing kerf-bent molding, avoid nailing through the kerfed portion as the nail can split the board.

STEAM BENDING A CURVED MOLDING

Steaming curved moldings requires that you build a steam box. Steam boxes can be anything from a length of downspout set over a teakettle to a propane fired boiler and wooden box system made exclusively for steaming.

Different species of woods steam-bend differently so it is best to experiment a bit before choosing an appropriate wood for your situation. Some woods that will bend easily in response to steam are: all the oaks, ash, fir, and cedar. Due to the tendency for steamed wood to spring back a bit when removed the piece will not come off the forms exactly true to the form. On larger dimension material (thicker than 1 in.) it may be necessary to bend the piece to a tighter radius to allow it to spring back a little.

Steambending guidelines

- You want to produce steam that is full of moisture. It should look dense and thick coming out of the box—not just wisps of white smoke. The more of it the better.
- Make sure the steam can move easily through the steam box. You absolutely do not want an airtight box.
- As a rule of thumb: Steam one hour per inch of material thickness.
- If a piece does not seem like it is going to take the required bend, immediately put it back in the box and let it steam longer.
- The last foot or so of a board is very difficult to bend. Make the piece longer than necessary and continue the bend further than necessary, leaving the excess on the ends straight.
- Hot wood coming out of the steam box is very soft so always use soft wood or plastic pads under the clamps to prevent damage.
- Never allow steel clamp heads to bear against the wood—most wood (and especially oak) will be deeply stained by the iron in steel.
- Do a trial run to make sure everything needed is on hand. There is only a small window of time to get the piece bent and you need to work quickly without interruption.
- Having an extra person will make the job go easier and faster. It is difficult to hold the bent board and clamp it in place at the same time.

Bending the molding

Once the steam is pouring thickly out of the steam box, insert the stock into the steamer and note the time ❶. At the appropriate time, remove the piece from the steamer and start the bending process over your knee on the way to the bending station ❷. Beginning at one end, clamp the piece to the first station on the bending form ❸. As you pull the piece around the curve, clamp it as you go. It is important to get the majority of the bend finished as soon as possible so if you find it is taking too long skip ahead to the opposite end and loosely clamp it in place ❹. Then quickly go back and clamp the remaining stations ❺. Once the clamps are all in place let the piece cool off for at least an hour or until it has set and remove the clamps.

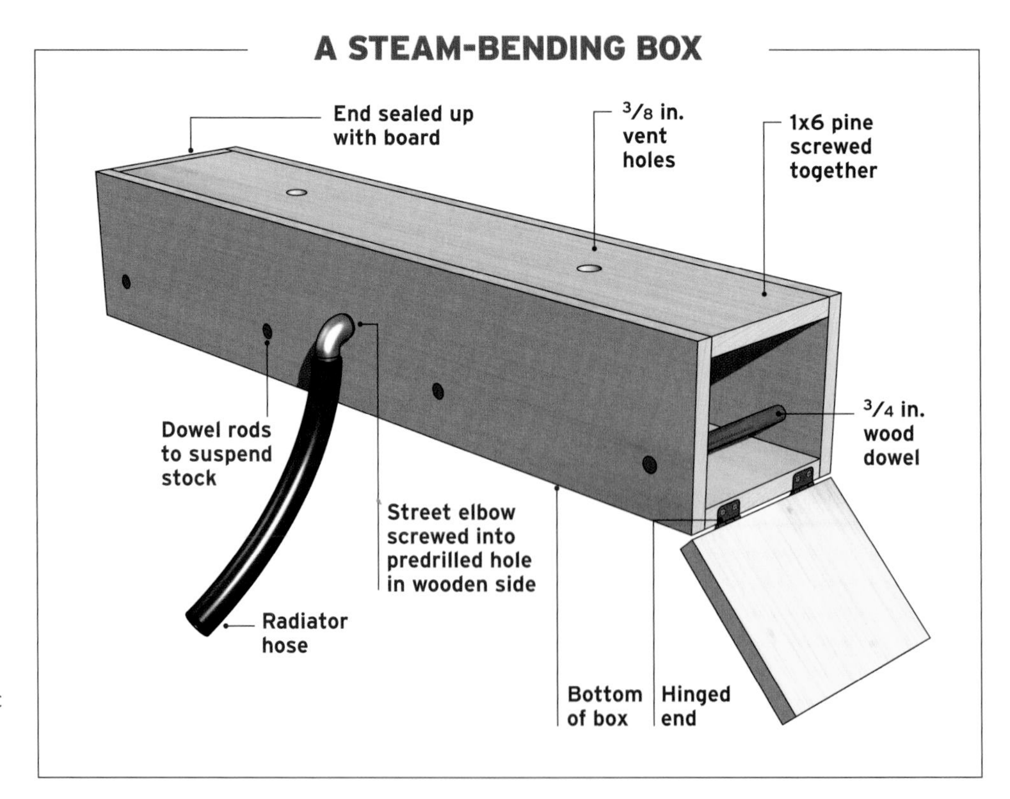

1 Once steam is up insert the piece into box and mark the time.

2 Remove from steamer and start the bend over your knee.

3 Clamp one end of the piece into the bending form.

4 Get the majority of the curve bent as soon as possible.

5 Quickly go back and install the remaining clamps.

WARNING Steam is extremely hot and can burn easily. Always wear appropriate clothing and gloves when working around steam. Also make sure the steam box has openings for the steam to escape easily to avoid explosive pressure buildup.

FRAME-AND-PANEL FAUX POST

When designing a faux wrap for a column keep in mind that they often look better if they have more mass at the bottom—an effect that makes them appear "grounded" to the floor. The corners of the wrap are either mitered or butt jointed. Properly constructed mitered corners showing no joint line give the wrap the appearance of being a single piece of wood. Butted corner joints are much easier to do but show joint lines, especially if there is an obvious grain change along the edges of the boards. Take the time to make a scale drawing to determine the proportions of all the post's design elements.

Start by applying the first layer of material to the sides of the post. This creates the face of the recessed panel above and serves as backer material for the base ❶. If desired, a lower grade of material may be used for the base area, as the base wrap will cover it. Now install the mitered lower column wraps over the lower section of the column, gluing and nailing the mitered corners together ❷. Next, install the mitered base cap molding to the column above the base plywood ❸. Now rip the stiles for the upper column to width with a 45-degree miter along one edge. Crosscut the stiles to length and nail them to the column, using a piece of mitered scrap as a positioning guide when nailing the first stile piece in place ❹. If a gap appears along the edge, trim the angle of the stile's edge miter accordingly. Now apply glue to both miter faces and nail the miters together and then to the backing panel ❺. Cut the top and bottom rails to length, apply glue to the butt ends and install between the stiles ❻. Now apply the cove transition moldings to the top and bottom of the base cap molding ❼ and install the cove moldings to the inside of the panel frames ❽. Install the crown molding ❾ and finish with the baseboard wrap ❿.

1 Cut base layer panels to size and nail material to the post.

2 Miter the corners and install the lower column wrap.

5 Glue and nail the corner stiles to each other and to panel.

6 Glue and nail top and bottom rail in position between the stiles.

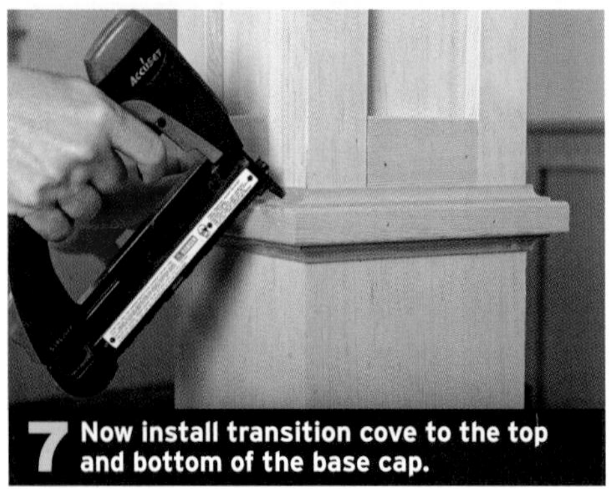

7 Now install transition cove to the top and bottom of the base cap.

3 Nail and glue the mitered base cap molding pieces to the post.

4 Use a piece of mitered scrap to align the first mitered corner stile.

8 Cut the molding and nail to the inside of the frame.

9 Install crown molding to top of column.

10 Install baseboard to bottom of column.

INSTALLING BUILT-UP CROWN MOLDING

Before ordering materials, draw your crown molding and determine the installation sequence. Generally, the flat moldings are installed first, followed by molded profiles and crown molding.

The ideal situation is to make an actual mock-up. Map out the splices and plan to minimize the visual impact of coped inside corners.

➔ **See "Planning Trim for Best Appearance," p. 3.**

Use color coded chalk lines to lay out the blocking (red) and (erasable blue) for the finished lines ❶. Rip 3/4 in. birch or 1/2 in. AC plywood into strips (width depending on size of blocking you will need) for nailing cleats. Apply construction adhesive and nail the cleats through the drywall into ceiling joists and wall studs ❷. Rip strips of plywood to create an L and nail it to the cleats ❸. Once the blocking is installed around the entire perimeter of the room, begin the installation sequence. Install flat moldings around the blocking first. Biscuit the butt joints to avoid the joints opening ❹. Install the profiled molding for the wall frieze, and then install the bed molding. Inside corners are coped, outside corners are mitered ❺. Miter and biscuit the ceiling frieze to ensure a tight joint. Glue and nail the frieze molding to the ceiling against the blue layout line ❻. The final step is to install the crown molding. Glue and nail across the miter to secure the joint ❼. At this point the installation is the same as installing simple crown.

➔ **See the section on Crown Molding, pp. 128–137.**

Work your way around the room, mitering outside corners and coping inside corners ❽.

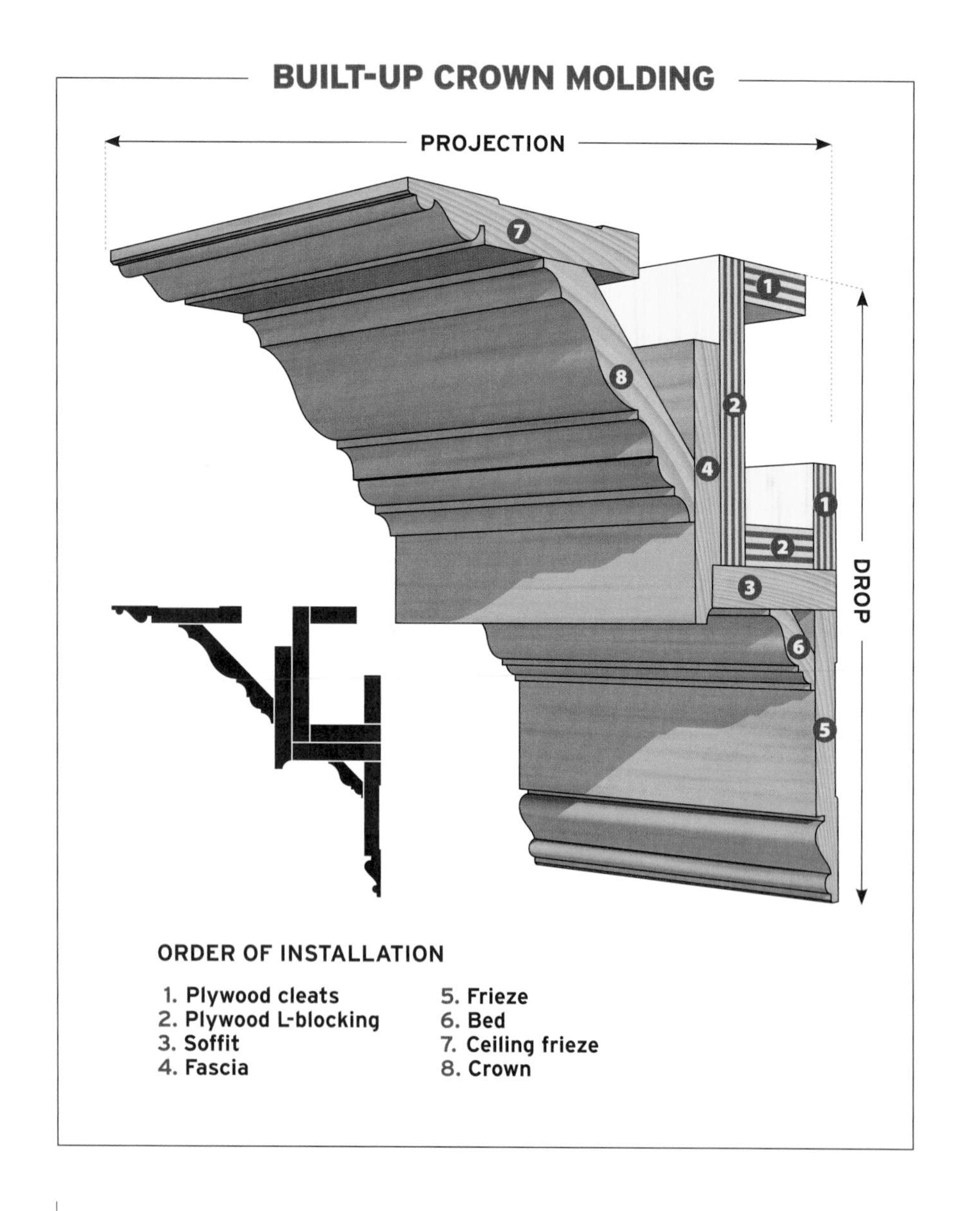

To support the ceiling frieze in a section between joists, cut a hole in the drywall inside the layout line, and screw a piece of plywood to the joists.

1 Measure the drop and projection and use color coded chalk lines to lay out blocking and finish marks.

2 Glue plywood cleats and nail through the drywall into ceiling joists and wall studs.

3 Nail two pieces of plywood together to create an L and nail to the cleats.

4 Biscuit the joints of the flat moldings and install the soffit and the fascia.

5 After installing the wall frieze, install the bed molding.

6 Install the frieze molding to the ceiling against the blue layout line.

7 Install the crown molding. Glue and nail across the miter.

8 Working your way around the room, install the remaining crown.

FIREPLACE SURROUND AND MANTEL

A fireplace surround and mantel can be simple or complex. This one uses frame-and-panel construction with plywood for the panels. For this project, begin by measuring the area to be covered by the surround and the size of the firebox. Draw a plan to these measurements. Begin by grooving the stiles and rails of the top panel and pilaster panels. Cut the rails on the tablesaw with a dado blade set to a slightly wider width than the thickness of the plywood ❶. The panel should fit snugly, but should slide into the frame easily. Take test cuts to get the proper fit. Cut the tenon cheeks on the tablesaw using a tenoning jig ❷. Clamp a stop block to the rip fence to position the shoulder cuts ❸. The fit should be tight, but nonbinding. Too tight will split the joint ❹. Dry fit all parts before assembly. Then glue up the top panel and the pilasters ❺. The pilasters have a short return. The outside edge abuts a column here, but you can also have returns on both sides of the pilaster. Miter the panel edges ❻ and one edge of each return. Cut biscuit slots in the miters of the panels and the returns ❼. Apply glue to the miters and in the biscuit slots, insert the biscuits and join the panel with the return. Use masking tape to hold together the miter ❽ or you can use light duty clamps to close any gaps. Once the panels are complete, add blocking to the pilasters and join them to the top panel with screws ❾. Add the molding to the top panel, using triangular plywood glue blocks to reinforce the top edge of the molding ❿. Install vertical blocking to hold the pilasters. In the case of masonry wall like this, use masonry screws to secure it ⓫. The finished surround and mantel completely hides the brick wall beneath ⓬.

1 Using a tablesaw outfitted with a dado blade, groove the rails and stiles.

2 Cut the tenon cheeks on the tablesaw using a tenoning jig.

6 Miter the sides of return and the panel and one edge of each return.

7 Cut biscuit slots in the miters of the panels and the returns.

9 Blocking reinforces the connection between the top panel and the pilaster.

10 Glue the molding to the top panel, using triangular plywood glue blocks.

3 Reset the height of the blade and cut the tenon shoulders.

4 A snug fit ensures a strong joint.

5 After a dry fit, glue up the top and pilaster panels.

8 Clamp the biscuited and glued miters with masking tape.

11 Install vertical blocking for the pilasters.

12 The finished surround and mantel.

GLOSSARY

APRON A piece of trim placed immediately below a window stool to cover any space in the rough opening.

ARCHITRAVE The lowermost part of an entablature resting directly on top of a column in classical architecture.

BASEBOARD A molding placed at the intersection of the wall and floor.

BEAD A narrow, half round molded profile applied to edges of boards.

BEVEL An angled cut applied at the edge of a panel or molding.

BISCUIT JOINT A method of joinery using precut elliptical mortises and wooden biscuits acting as tenons.

BOX NAILER A three- or four-sided hollow assembly that is attached to a ceiling and is used to create a faux beam.

BLOCKING Boards put in between existing structural framing to create additional backing for nailing.

BUTT JOINT A non-structural joint where the squared off edges or ends of stock are butted together. Requires fasteners or glue for strength.

CENTERING BATTEN A batten used to locate the center point on frequently occurring widths.

CHAIR RAIL Horizontally applied wall molding located at the approximate height of a chair back.

CLOSURE BOARD The last board installed when using boards to cover a wall or ceiling. Usually requires ripping or cutting to fit.

COPED JOINT A joint used on molded millwork for ninety degree inside corners where a matching profile is cut on the end of a molding allowing it to fit the existing piece.

CORNICE An assembly of moldings at the juncture of the ceiling and wall.

CROSS CUT When a board is cut across the direction of the grain.

CROSS-LEGGED OPENING Applies to rough openings for doors and occurs when the framing on one side of the door is not in vertical alignment with the other side.

CROWN MOLDING Molding applied at the intersection of the wall and ceiling.

END GRAIN A cross section or whenever a cross section of the wood shows.

ENTABLATURE Any raised horizontal architectural member.

EXTENSION JAMB The part of a window or door assembly that fills any space left showing of the rough framing.

FAUX BEAMS Non-structural beams.

FAUX POSTS Non-structural posts.

FRAME AND PANEL A type of construction consisting of a panel surrounded by a frame. Traditionally the panel was tapered along its edges and inserted into a mortise running along the inside of the frame.

HINGE BIND Refers to a condition where a door does not shut without some pressure due to the hinges being mortised incorrectly.

JAMB The outside framework of a door or window that covers the rough framing and in the case of a door provides a surface on which to mount the hinges.

JOINT The junction where two pieces of material are joined together.

MITER JOINT A joint consisting of two pieces cut at an angle to join each other.

MORTISE A hole cut in a wooden part meant to accept a tenon or another piece of wood.

MULLION A piece of molding used to cover the joint or space between two adjacent windows.

PARTING BEAD Narrow piece of molding used to separate different pieces of trim.

PINCH STICKS Two sticks of the same size used to determine distance between two points without measuring. The sticks are slipped past each other until they fit the opening and then are either marked or held in place until the measurement is transferred.

PITCH BREAK When a ceiling surface changes pitch or angle.

PLINTH BLOCK A trim element typically placed at the bottom of vertical door casings. Used to provide a transition between the vertical casing and the baseboard.

PREACHER A piece of wood notched to fit over the baseboard and held tight against the standing molding. Used for marking baseboard to correct length and cut angle.

PRE-HUNG DOOR A door that comes hung on its hinges in a jamb assembly.

PROFILE When an object is viewed from the side.

PROPORTIONAL DIVIDER A tool that divides an angle into two equal parts.

RABBET A square or straight-sided groove cut along the edge of a board.

RAIL The horizontal element of a frame.

RIP The process of cutting wood with the grain.

ROUGH OPENING Space left in wall framing to accommodate windows or doors.

SCRIBING The process where the profile of an irregular surface is transferred to a board prior to cutting.

SPILING BATTEN A thin piece of wood used to record and transfer the profile of an irregular surface.

SPRING ANGLE Refers to the angle at which crown molding will stick out into a room. Measured between the back face of the crown molding and the wall.

STILE The vertical element of a frame.

STOOL The wide horizontal board at the base of a window frame.

TAPERED RABBET A rabbet where one side is long and cut at a shallow angle creating clearance space and creating a shoulder for a fastener to land on. Used primarily when attaching extension jambs.

TENON A projection of wood on the end of a piece of wood meant to be inserted into a mortise.

TONGUE-AND-GROOVE A type of joint using a machined groove on one edge of a board and a tongue on the other. When adjacent boards are pushed together the tongue fits inside the adjacent groove.

UNDERCUT A less than ninety degree cut on the edge of a board allowing the face to fit tight against an adjacent surface.

WAINSCOT Decorative panels or boards installed on the lower portion of a wall.

INDEX